Tackling Environmental Health Inequalities in a South African City?

South Africa is widely recognised as a middle-income, industrialised nation, but it also ranks amongst the most unequal countries in the world in terms of its income distribution and human development. Environmental health remains a considerable public health challenge in the 21st century as Environmental Health Practitioners (EHPs) try to tackle local environmental health inequalities in the face of historically disadvantaged populations suspicious of their motives and demands that far exceed any resources available.

Based on an empirical research project that explores how local government Environmental Health Practitioners regulate environmental health in one of South Africa's largest, fastest growing and most unequal cities, Urbington, this book explores the many influences on their decision-making including the limits of the law, organisational controls, the views of EHPs themselves and their relations with businesses, communities, politicians and others.

Tackling Environmental Health Inequalities in a South African City? argues that if we are to meet the environmental health challenges of the 21st century, it is in our best interests to rediscover this vital local public health workforce. This book is essential reading for students, practitioners and policymakers in environmental health and public health, as well as those interested in urban development and policy, particularly in African cities.

Dr Rob Couch is a registered Environmental Health Practitioner with experience in the public, private, academic and charity sectors across the UK and internationally. He currently works in a shared public health team across three local authorities in the East of England. When not in practice his wider research interests include the environmental health workforce and how Doughnut Economics can help to create more equitable, inclusive and sustainable cities. Rob is also a co-founder of the UK Environmental Health Research Network, which promotes the role of research, publication and more evidence-based environmental health.

Routledge Focus on Environmental Health

Series Editor: Stephen Battersby, MBE PhD, FCIEH, FRSPH

Statutory Nuisance and Residential Property
Environmental Health Problems in Housing
Stephen Battersby and John Pointing

Housing, Health and Well-Being
Stephen Battersby, Véronique Ezratty and David Ormandy

Selective Licensing
The Basis for a Collaborative Approach to Addressing Health Inequalities
Paul Oatt

Assessing Public Health Needs in a Lower Middle Income Country
Sarah Ruel-Bergeron, Jimi Patel, Riksum Kazi and Charlotte Burch

Fire Safety in Residential Property
A Practical Approach for Environmental Health
Richard Lord

COVID-19: The Global Environmental Health Experience
Chris Day

Regulating the Privately Rented Housing Sector
Evidence into Practice
Edited by Jill Stewart and Russell Moffatt

Dampness in Dwellings
Causes and Effects
David Ormandy, Véronique Ezratty and Stephen Battersby

Tackling Environmental Health Inequalities in a South African City?
Rediscovering Regulation, Local Government and its Environmental Health Practitioners
Rob Couch

Tackling Environmental Health Inequalities in a South African City?

Rediscovering Regulation, Local Government and its Environmental Health Practitioners

Rob Couch

LONDON AND NEW YORK

First published 2023
by Routledge
4 Park Square, Milton Park, Abingdon, Oxon OX14 4RN

and by Routledge
605 Third Avenue, New York, NY 10158

Routledge is an imprint of the Taylor & Francis Group, an informa business

British Library Cataloguing-in-Publication Data
A catalogue record for this book is available from the British Library

ISBN: 978-0-367-44468-6 (hbk)
ISBN: 978-1-032-52986-8 (pbk)
ISBN: 978-1-003-00993-1 (ebk)

DOI: 10.1201/9781003009931

Typeset in Times New Roman
by MPS Limited, Dehradun

For Ella and Freya

For all those working to tackle environmental health inequalities, with thanks

Contents

Figures and Tables

Figures

Tables

Series Preface

This is the 12th publication in the Routledge Focus on Environmental Health series, with more in the pipeline. This edition reflects our desire to highlight environmental health work around the world. There are lessons that environmental health practitioners can learn from colleagues working in different settings and in a variety of social, political and legal structures. Yet fundamentally the aim is the same, to protect public health, that is prevent ill-health and unintentional injuries.

It further illustrates the flexibility offered by the series, but the aim remains as ever; to explore environmental health topics traditional or new, and raise sometimes contentious issues in more detail than might be found in the usual environmental health texts. It is a means whereby environmental health issues can be discussed with a wider audience in mind.

This series is an important part of the professional landscape, as is apparent from the titles published so far. Environmental health practitioners bring their expertise to a range of situations and are deployed in a variety of ways, but not always to the best effect so far as public health is concerned. All too often politicians both at the national and local levels are unaware of what environmental health is, what practitioners do or how they work. It is common that practitioners have a 'low profile' or are taken for granted. It is hoped that this series will aid in highlighting the work of environmental health practitioners.

We want to encourage readers and practitioners, particularly those who might not have had work published previously, to submit proposals as we hope to be responsive to the needs of environmental and public health practitioners. I am particularly keen that this series is seen as an opportunity for first-time authors and as ever would urge students (whether at first- or second-degree level) to consider this an avenue for publishing findings from their research. Why for example should the hard work that has gone into a dissertation or thesis lie unread on a library shelf? We can provide advice on turning a thesis into a book. Equally this series can be a way of extending a presentation, paper or training materials so that these can reach a wider audience.

The series provides a route for practitioners to improve the profile of the profession as well as provide a source of information. It has the advantage of having a relatively quick turnaround from submission of the manuscript to publication and can thus be more up-to-date and immediate than a standard textbook or reference work.

EHPs have perhaps fallen short in telling others about their work. To garner more respect for the profession and to develop professionally, we encourage EHPs on the front line to “get published”, writing up their work of protecting public health. This is a route for analysing actions and reporting on what worked in practice, what was successful, what wasn’t – and why. Useful insights for others working in the field can be provided and also highlight policy issues of relevance to environmental health.

Contributing to this series should not be seen merely as an exercise in gathering CPD hours, but as a useful method of reflection and aid to career development, something that anyone who considers themselves a professional should do. I am pleased to be working with Routledge to provide this opportunity for practitioners.

As has been made clear it is not intended that this series takes a wholly “technical” approach but provides an opportunity to consider areas of practice in a different way, for example looking at the social and political aspects of environmental health in addition to a more discursive approach on specialist areas.

Our hope remains that this is a dynamic series, facilitating a forum for new ideas and debate on environmental health topics. If readers have any ideas for titles in the series please be encouraged to submit them to me as series editor via the e-mail address below.

“Environmental health” can be taken to mean different things in different countries around the World and so we welcome suggestions from a range of professionals doing “environmental health” work or policy development. EHPs may be a key part of the public health workforce wherever they practise, but there are also many other practitioners working to safeguard public and environmental health. It is our hope that this series will enable a wider range of practitioners and others with a professional interest to access information and also to write about issues relevant to them.

Forthcoming titles are likely to cover such topics as air pollution and sewage pollution. We are in contact with colleagues around the world encouraging them to submit proposals. That does not mean we have no need for further suggestions, quite the contrary, so I hope readers with ideas will get in touch via Ed.Needle@tandf.co.uk.

Series Editor: Stephen Battersby
MBE PhD, FCIEH, FRSPH

Acknowledgements

My thanks to Professor Trudy Harpham, Dr Adrian Budd and the late Professor Michal Lyons for their encouragement, support and guidance during and since my time studying with the Urban Development and Policy Group at London South Bank University.

All thanks to friends and former colleagues at the Medical Research Council of South Africa for sharing their knowledge and making me feel at home in South Africa. I am also very grateful for the advice and guidance of so many academics and EHPs from across South Africa and the UK.

Finally, my thanks to all those of the Department of Health of Urbington Metropolitan Municipality for tolerating my presence and giving me their valuable time.

1 Introduction

> Everyone has the right to an environment that is not harmful to their health or well-being ...
>
> (Section 24, Constitution of the Republic of South Africa, 1996)

> Shops are falling apart and people can't fix them because they don't get enough income. So you go there, you inspect and you give recommendations and somebody doesn't have money to even paint!
>
> (Urbington Township EHP)

Local government Environmental Health Practitioners (EHPs) have been regulating environmental health in South African cities for more than 100 years. Based at the local level they are now well placed to bring protective laws, including the constitutional right to environmental health, to life. Indeed, their work in prevention remains key to delivering Professor Michael Marmot's challenge:

> Why treat people and send them back to the conditions that made them sick?
>
> (Marmot 2015:1)

However, the work of local government EHPs remains largely invisible to the public, politicians and policymakers, whilst EHPs themselves have to grapple with dilemmas like those of the Township shop above each and every day. If we're to meet the public health challenges of the 21st century, I think it's in our best interests to remove this cloak of invisibility, to rediscover what EHPs do, how they do it and their relevance for creating and maintaining the inclusive and sustainable cities we urgently need.

Local government EHPs have to be registered with the Health Professions Council of South Africa to practice. Unlike many other

DOI: 10.1201/9781003009931-1

health professionals, their work tends to be upstream, focused on tackling what Professor Marmot calls the causes of the causes of inequalities, in contrast to downstream interventions that seek to reduce harm instead of preventing it (Dhesi 2019). This work is largely regulatory and could be considered a form of social regulation in the public interest to protect against the damaging consequences of market failure like unsafe food or housing (Ogus 2004). But my findings also reject simplistic readings of regulation as the triumph of public interests, here local government EHPs, over private capital. Instead, this book paints a more complex picture where power circulates continuously between many actors and is distributed unequally and is sometimes not with EHPs themselves, but spaces for challenging it are plentiful and rarely closed down.

This book is based on an in-depth study of ten front-line EHPs working across Urbington, a city municipality chosen to represent the diversity of urban environmental health across South Africa. Over 6 months in 2007, EHPs working in homes, businesses, schools, etc. were observed in their daily practice and then interviewed alongside an analysis of their paperwork. Observations and interviews were also conducted with other EHPs and their managers, together with an analysis of policy documents. A model was developed to summarise how EHPs try to tackle environmental health inequalities, the word 'try' explains why the main title of this book ends with a question mark. Measuring whether inequalities are tackled or not remains challenging but strengths and weaknesses are identified and recommendations are made towards more effective and visible working.

The accounts and quotations below sometimes make reference to racial identities that are left unedited because EHPs still use race to describe their worlds and navigate them, I apologise for any offence caused. Any emphasis in the quotes of EHPs is also their own which I've tried to convey using block capitals. My findings also pre-date the COVID-19 pandemic and the vital work of EHPs in response but I direct readers to Day et al. (2021), Morse et al. (2020), Rodrigues et al. (2021) and Zimba (2022).

In Chapter 2, I set the scene by considering why local government EHPs matter. I start by considering the relevance of their work by exploring South Africa's environmental health challenges. I then introduce Urbington Metropolitan Municipality's response to these challenges, including a profile of its EHPs and their typical roles and working days. Chapter 3 introduces the model of how they try to tackle environmental health inequalities. I start by rediscovering the concept of responsive regulation and how Urbington EHPs use it when deciding whether to punish or persuade offenders. I then explore the law enforcement and project/promotion pathways they use before embedding these within a wider framework of governance to capture the many factors influencing their decisions.

Chapter 4 is the first to unpack this governance model and explore in more detail why persuasion is frequently favoured by EHPs over punishment. It begins by looking at the regulatory context, particularly how apartheid, macro-economic policy and public and media attitudes shape the work of Urbington's EHPs. It then explores the law and how its great breadth, uncertainties and limits create many challenges for the EHPs responsible for bringing it to life in the streets of Urbington. Chapter 5 focuses on the municipality itself and starts with its vision and the work of EHPs in monitoring conditions for others. Vital relationships between EHPs, their managers and politicians are then considered before a review of the very limited resources for environmental health services.

Chapter 6 continues to unpack the governance model by focusing on what EHPs themselves think. It starts by discussing their motivations and very real safety concerns before exploring their views on law enforcement and troubles with sanctions and the criminal courts. Chapter 7 considers relations between EHPs and the regulated and starts by exploring how EHPs categorise those they encounter before looking at other characteristics including ethnicity and morality. The chapter ends with focused sections on the challenges of the informal sector and corruption.

The final chapter pulls together my findings, revisits key questions and makes recommendations to strengthen the work of local government EHPs. I hope my findings and discussions will resonate with all EHPs and others facing similar challenges. Environmental health inequalities are a global problem, we have much to learn from each about how to tackle them and we are running out of time.

References

Day, C., R. Couch and S. Dhesi (2021) *Covid19: The Global Environmental Health Experience*. Routledge, London, UK.

Dhesi, S. (2019) *Tacking Health Inequalities: Reinventing the Role of Environmental Health*. Routledge, London, UK.

Marmot, M. (2015) *The Health Gap*. Bloomsbury, London.

Morse T., K. Chidziwisano, D. Musoke, T. Beattie and S. Mudaly (2020) Environmental health practitioners: a key cadre in the control of COVID-19 in sub-Saharan Africa. *BMJ Global Health*, (5).

Ogus, A. (2004) *Regulation: Legal Form and Economic Theory*. Hart Publishing, Oxford, UK.

Rodrigues, M. A., M. V. Silva, N. A. Errett, G. Davis, Z. Lynch, S. Dhesi, T. Hannelly, G. Mitchell, D. Dyjack and K. E. Ross (2021) How can Environmental Health Practitioners contribute to ensure population safety and health during the Covid-19 pandemic? *Safety Science*, Vol. 136 (5).

Zimba, A. (2022) Why is it worth it? Adaptation and resilience of Environmental Health Practitioners in implementing key intervention strategies in the Covid-19 pandemic. *Environment and Health International* Vol. 22 (1) 19–26.

2 Why local government EHPs matter?

Environmental health inequalities: a wicked and doughnut-shaped challenge

The term 'environmental health' has been in common use for at least 50 years but remains hard to define (CEH 1997; Smith et al. 1999) for many reasons, not least because 'environment' and 'health' are themselves hard to define (Eyles 1997). This book uses the definition developed in 1994 by the World Health Organisation and later adopted by the South African government (NDoH 2013):

> Environmental health comprises those aspects of human health, including quality of life, that are determined by physical, chemical, biological, social and psychosocial factors in the environment. It also refers to the theory and practice of assessing, correcting, controlling and preventing those factors in the environment that can potentially affect adversely the health of present and future generations.
>
> (WHO in MacArthur and Bonnefoy 1997:5–6)

South Africa's environmental health inequalities can be described, after Rittel and Webber (1973), as a wicked problem because their unique and complex nature defies easy definition, whilst policy responses like the work of local government EHPs remain hard to describe and test. For centuries, cities have symbolised both civilisation and wealth and poverty and deprivation (Chadwick 1842/1965; Rosen 1993). Concerns about environmental health inequalities and their impacts on city dwellers continue (Hardoy et al. 2001; Harpham et al. 1988; UN 1976; WHO 2008) and are increasingly relevant now

DOI: 10.1201/9781003009931-2

more than half the world's people are urban and the populations of African and Asian cities are predicted to double between 2000 and 2030 and be largely composed of the poor (UNFPA 2007).

For more than 30 years these inequalities in the social determinants of health have been visualized using Dahlgren and Whitehead's (1991) rainbow model which maps the relationships between the individual, their environment and health. Dahlgren and Whitehead (2021) recognise its continued strengths including its engagement with the upstream socio-economic influences on health that are holistic, interconnected and multisectoral but amenable to organised action by society, in contrast to more disease-focused models of health and 'downstream' health services related interventions. Here, the actions of local government EHPs (and others) in assessing, correcting, controlling and preventing the potentially harmful social determinants that shape peoples' living and working conditions remain inherently upstream and relevant.

The rainbow model is not reproduced here because Raworth's (2017) Doughnut economics model for the Anthropocene in Figure 2.1 updates it for the 21st century. It places the rainbow within the limits of our finite planet, whilst acknowledging the role of human activity in creating these challenges and in finding a way out of them. The dark circles depict the boundaries of the doughnut itself, the ecologically safe and socially just space in which all humanity can thrive. The inner boundary represents the social foundation, below which shortfalls are experienced across twelve dimensions based on the United Nation's 2015 Sustainable Development Goals. Here the constituent parts of Dahlgren and Whitehead's (1991) rainbow co-exist with other vital aspects including political voice, social equity and gender equality that remain integral to tackling environmental health inequalities.

The Doughnut's outer boundary represents an ecological ceiling based on Steffen et al's (2015) planetary boundaries framework, any overshoots across its nine dimensions potentially endanger Earth's life-supporting systems. These include more familiar challenges for EHPs like air and chemical pollution and climate change alongside those beyond their scopes like biodiversity loss and ocean acidification. The dark-shaded wedges in Figure 2.1 also attempt to quantify the best available evidence of the pressures on shortfalls/overshoots in 2017. Chapter 8 revisits why local government EHPs could be viewed as doughnut makers, but I now explore environmental health inequalities in more detail.

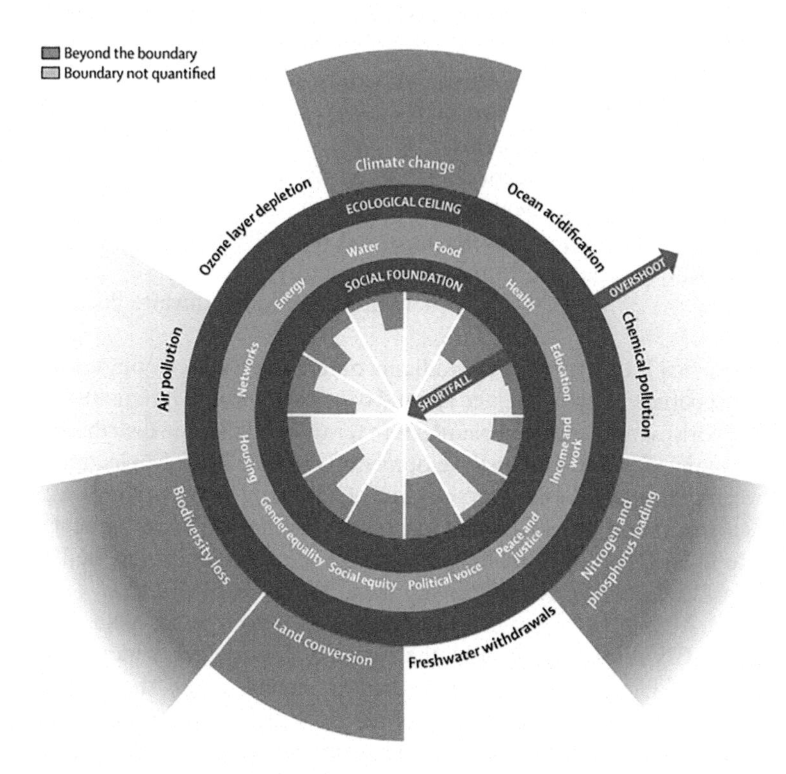

Figure 2.1 Raworth's (2017) doughnut for the Anthropocene.

South Africa is affected by three concurrent epidemics (Coovadia et al. 2009); poverty-related ill health remains widespread, whilst HIV/AIDS accounts for 31% of disability-adjusted life years and the burdens from non-communicable diseases and other causes, particularly violence and road traffic injuries, are increasing. To explore this further Mitlin and Satterthwaite's (2004) different aspects of poverty are useful here because of their foundations in the complex and interlinked social determinants of health, as captured by Raworth's (2017) doughnut, where one aspect may cause another but addressing one could resolve or lessen others. These different aspects include:

- Inadequate and often unstable income;
- Limited or no safety net (e.g. access to grants);
- Inadequate, unstable or risky asset base (e.g. housing);

- Poor quality and often insecure, hazardous and overcrowded housing;
- Inadequate provision of public infrastructure;
- Inadequate provision of basic services (e.g. day care, transport, law enforcement);
- Inadequate protection of poorer groups' legal rights (e.g. to environmental health);
- Poorer groups' voicelessness and powerlessness within political systems and bureaucratic structures

(Mitlin and Satterthwaite 2004:15)

South Africa is described as a middle-income, industrialised country but its Gini coefficient has remained around 0.7 since 2000 and it remains one of the world's most unequal countries in terms of its income distribution (Sulla et al. 2022) and human development (UNDP 2015a). For example, between 1980 and 2014 the average years of schooling increased by 5 years (to 9.9 years) and gross national income per capita increased by nearly 12% to $12,122 (in 2011) but life expectancy at birth has only increased by 0.5 years and only started improving from 2005 (UNDP 2015b).

South African cities provide more jobs and higher wages than rural areas (SACN 2004) but in 2011 the national Census found more than one-third of Urbington's households had an income of less than R1633 (~$255) per month and one-quarter of its economically active population were unemployed. Small, medium and micro-enterprises[1] (SMMEs) are estimated to employ around 2 million across South Africa (DTI 2008) but many have little or no employment protection. An emerging social welfare system has seen more than 16 million South Africans earning less than R6000 ($826 in 2011) per household per month benefit from interventions including free primary health care, no-fee schools, social grants for vulnerable groups, free/subsidised housing and free basic services (National Treasury 2013) but many environmental health challenges remain.

Urban poverty still broadly follows the spatial patterns of apartheid with former Black-African areas experiencing the poorest access to basic services whilst former White suburbs still enjoy the highest standards and benefit from the relocation and growth of former inner-city manufacturing and service sectors (SACN 2004). Certain groups also have greater environmental health vulnerabilities, particularly women and children, those affected by HIV/AIDS (Hardoy et al. 2001; WHO 2008; Norman et al. 2010) and migrants (SACN 2004). Many South Africans also face the double burden from exposures to both traditional and modern (Thomas et al. 2002) or brown and green hazards (McGranahan and Satterthwaite 2000) or shortfalls and overshoots (Raworth 2017).

Shortfalls include inadequate access to basic services like housing, water and sanitation for reasons including historic backlogs in the state housing programme, population growth and the 'weak rights' of informal households, whose frequently 'illegal' status can make them even more vulnerable to exploitation and powerless to hold officials to account (Huchzermeyer 2004). Indeed, the estimated total burden of disease in South Africa attributable to unsafe water and sanitation, indoor and outdoor air pollution and lead exposure was around 5% of total deaths, though for under 5-year-olds it was nearly 11% (Norman et al. 2010). For overshoots, air pollution challenges include the spatial geography of apartheid cities that necessitated long journeys to work on polluting public transport and the continued use of solid fuel or paraffin burners in townships due to shortfalls in electrification and high electricity prices (SACN 2004).

Introducing Urbington and its EHPs

There is little published on local government EHPs in South Africa before democracy but brief mentions suggest work characterised by law enforcement that increased environmental health inequalities. The earliest records date back to the mid-17th century when European Cape settlers enacted and enforced laws including those to protect drinking water from contamination (Rabie and Fuggle 1992). From 1795 the British gradually created racially based local authorities across Southern Africa (de Visser 2005). Colonial sanitary inspectors enforced the racial segregation of urban areas (Parnell 1993) in response to outbreaks of disease (Swanson 1977) and the belief that segregation could increase labour controls and solve problems like insanitary housing and prostitution (Mabin 1986). This 'sanitation syndrome' equated Black urban settlement, labour and living conditions with threats to public health whilst rationalizing White fears and prejudices and the need for official social controls (Swanson 1977). Colonial local authorities also extended their regulation of the urban space to control African street vendors (Parnell 2002; Rogerson 1986) but remained reluctant to provide basic services to emerging townships (Proctor 1986) and urban slums (Parnell 1991).

After independence in 1948, the environmental health of most South Africans continued to deteriorate during apartheid. The environment was a low national priority next to industrialisation and economic development, particularly from the 1970s when ongoing sanctions, boycotts and recession prioritised economic survival (Steyn 2008). Apartheid also reinforced racial local government structures (de Visser 2005), with cities run by Whites for Whites (Beall et al. 2000) and able to sustain the highest standards thanks to substantial revenues and Black migrant labour (de Visser 2005). However Black areas had separate local government structures or none at all, whilst apartheid limitations on businesses produced inadequate revenues for service provision (RSA 1998). Without basic services, these areas were prone to disease outbreaks that provided renewed justification for the removal of Black-Africans from urban areas (Andersson and Marks 1988; Ferrinho et al. 1991).

Regulators also failed to protect environmental health because of fragmented responsibilities, poor resources (Petrie et al 1992), a lack of political will (Davies et al. 1997; Rabie and Fuggle 1992) and the exemption of large, state-owned, polluters from the law (Steyn 2008). From the mid-1980s attempts were made to restructure local government and direct funding towards priority areas like the townships (RSA 1998), but in 1994 the democratic government inherited approximately 12 million people without access to clean drinking water, 21 million with inadequate sanitation (ANC 1994) and a local government characterised by racism, weakness and illegitimacy (de Visser 2005).

Local government EHPs, therefore, find themselves in a challenging context for the 21st century. As well as facing the wicked and doughnut-shaped inequalities outlined earlier, the scale of the transformation and restructuring of South African local government is daunting and historically disadvantaged populations are understandably suspicious of their motives, whilst the pressure is on them to maintain the highest standards in former White areas. Pressures to change their roles are also significant. For example, though EHPs have been described as the 'general practitioners' of public health (Cornell 1996) others are concerned they remain stuck in their law enforcement roles and unwilling to embrace more flexible, multidisciplinary, community development approaches (Burke et al. 2002; CEH 1997; Dhesi and Lynch 2016; Eales et al. 2002). McGranahan (2007) also argues that EHPs should not see their roles as inspecting and regulating but as working with communities and others improving conditions in cities. But others have found South African EHPs remained uncertain of (Allison 2002; Lewin et al. 1998), even hostile towards (McDonald 1997) these new roles.

Urbington local government has been through continual restructuring since 1990 and in 2007 its structure, summarized in Figure 2.2, provided politicians with ultimate legislative and executive power via both directly elected Ward Councillors, with whom most EHPs were in regular contact, and Councillors elected by proportional representation. A Speaker heads the legislature and oversees all committees whilst a Mayor is elected every 5 years and supported by a

Mayoral Committee whose members have specific portfolio responsibilities. For example, Environmental Health, as a sub-directorate of the Department of Health, was led by the Member of the Mayoral Committee for Health (hereafter MMC:Health).

The structure of Urbington itself was characterized by the division of policy-making from implementation in a so-called client-contractor split. Here the core 'client' administration was led by a City Manager who was also responsible for policymaking in areas including finance, planning and contract management. Central departments including Health, where EHPs were based, were also part of this core administration and responsible for policy making and implementation. Contractors comprised the regional administrations and utilities, agencies and corporatized entities. Each region acted as a contractor to its central department, with services like environmental health responsible for meeting performance targets, expected to liaise with Ward Councillors and be responsive to local citizens. The utilities, agencies and corporatized entities are owned by Urbington but intended to function as arms-length companies and EHPs often worked closely with them in areas including housing, water, sanitation and solid waste management.

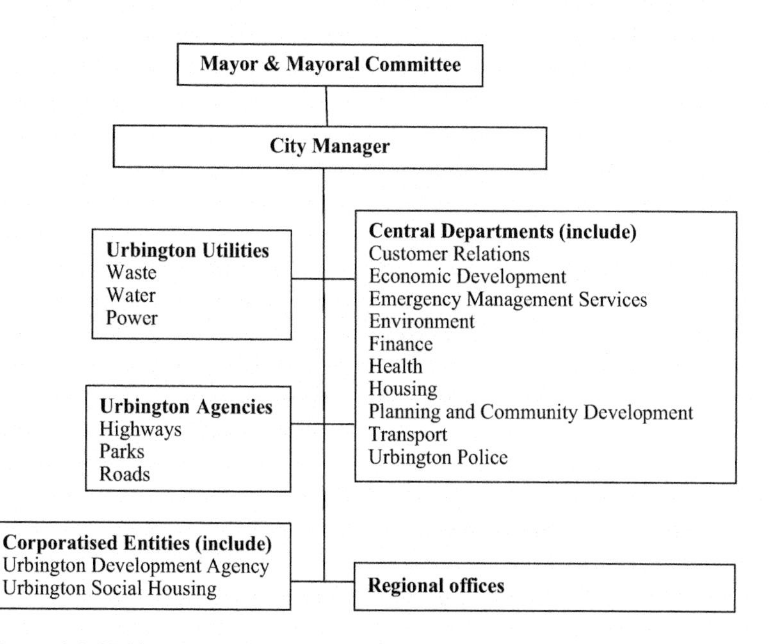

Figure 2.2 Urbington metropolitan municipality in 2007.

The structure of the Department of Health is summarised in Figure 2.3 and was led by the MMC: Health Councillor and a Director of Health and employed more than 1000 people across five sub-directorates in central and regional offices and clinics working towards:

> Vision: A City with a high quality, efficient, accessible and equitable health system for all, that has adequate and flexible capacity to meet the changing health challenges facing Urbington.
>
> Mission: Improved general health, well-being and increased life expectancy of the citizens of Urbington

In 2007 the responsibilities of the five sub-directorates included:

- Resources – Cross-department support, policy-making and management advice;
- HIV/AIDS – Coordinating HIV/AIDS-related programmes and projects;
- Public Health – Surveillance of communicable and non-communicable diseases and health data management;
- Primary Health Care – Delivery of primary health care services;
- Environmental Health – Complaints investigation, compliance monitoring, law enforcement and health promotion and training.

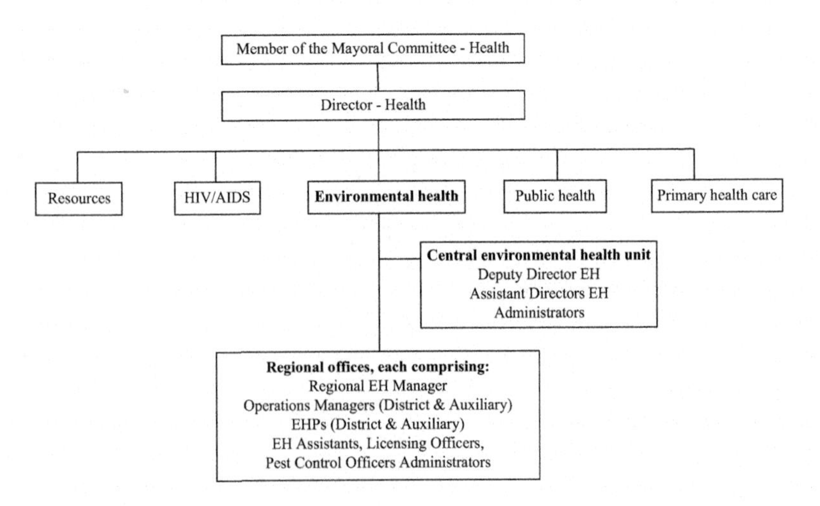

Figure 2.3 Urbington's department of health in 2007.

The Environmental Health sub-directorate comprised a Central Unit and more than 100 EHPs working across its regional offices. The Central Unit was led by a Deputy Director and a team of Assistant Directors (all experienced EHPs), supported by a data analyst and secretary and responsible for overseeing and supporting the regions and developing policy.

Until around 2004 EHPs had operated alone from clinics that were closer to communities in theory, but in practice EHPs found themselves scattered across Urbington and often unreachable because of this and a lack of basic facilities like telephones. EHPs now valued working together with colleagues and managers from regional offices and with more resources including administrators to receive complaints or answer questions. Of the four regional offices visited, two inner city ones occupied one open-plan floor alongside other municipal services. EHPs in the suburban and township offices had their own offices within larger municipal compounds. Most EHPs had their own desks, computers, telephones and filing cabinets within open-plan offices, though a few suburban EHPs still had their own offices. A strong feeling of office camaraderie was observed, EHPs clearly valuing being together to share information, knowledge and gossip. A few expressed concern about the lack of privacy in open-plan settings but other rooms were available nearby for meetings etc. Managers had their own offices nearby which also enabled them to observe their EHPs.

A profile of the Urbington EHPs observed in depth is summarised in Table 2.1. In contrast to previous studies (Hutter 1988; Mathee et al. 1999) male EHPs no longer dominated local government, with around half female and comprising mostly Black-Africans in their 20s–30s, with a handful of Coloured, Indian/Asian and White EHPs of a similar age. A later workshop with nearly all of Urbington's EHPs suggested this was typical of the whole workforce. Around half of Operations Manager and Regional manager posts were occupied by Black-African EHPs, with the rest White.

A Black-African background conferred some advantages in building relations with Black-African businesses and the public, particularly when languages, cultures and histories were shared. One White female EHP also thought her race and gender together put her at a greater safety risk. Half the EHPs were raised in rural areas in different Provinces, but four were born and raised in Urbington and most lived within the city itself though in different regions to those they worked in. Some managers thought ex-rural EHPs initially found it hard to adapt to Urbington's greater emphasis on law enforcement but, unlike

Table 2.1 A profile of Urbington EHPs

EHP	*Sex*	*Age*	*Race*	*Area of upbringing*	*EH education*	*Pre-Urbington EH experience (years)*	*Urbington EH experience (years)*
Suburban EHP 1	M	24	Black-African	Rural	Diploma; B.Tech.	Provincial Gov - 1.5 National Gov - 1	0.25
Suburban EHP 2	M	29	Black-African	Rural	Diploma; B.Tech.	Provincial Gov - 1	5
Suburban EHP 3	M	59	Black-African	Township	Diploma	None	32
Township EHP 1	F	34	Black-African	Township	Diploma	None	8
Township EHP 2	F	32	Black-African	Township	Diploma; B.Tech.	Provincial Gov - 4	2.5
Inner City EHP 1	M	24	Black-African	Rural	Diploma; B.Tech.	Provincial Gov - 1 National Gov - 0.3	2
Inner city EHP 2	M	35	Black-African	Rural	Diploma	Local Gov - 1	5
Inner city EHP 3	F	31	Black-African	Township	Diploma; B.Tech.	Provincial Gov - 1 Private Sector - 6	2
Inner city EHP 4	M	33	Black-African	Rural	Diploma; B.Tech.	Private Sector - 1 Local Gov - 6	3
Inner city EHP 5	F	44	White	Urban	Diploma	None	12

Hutter (1988), no ex-rural EHPs observed seemed reluctant to use more punitive approaches.

All EHPs had National Diplomas in Environmental Health and were registered with the Health Professions Council of South Africa, as required by the law to practice as an EHP in South Africa. Most had also completed or were near completing their Bachelor of Technology degrees in Environmental Health Sciences and were interested in continuing education to develop their specialist knowledge (e.g. Certificates in Noise Control or Safety Management, Masters in Environmental Health or Solid Waste Management) or management skills (e.g. Business Studies, Public Administration, Financial Management). Some also considered local government experience crucial to their career development:

> … to really practice being an Environmental Health Practitioner you have to do it at the local level, at the municipality.
>
> (Inner city EHP 2)

This was partly because the breadth of local government work wasn't available in other sectors, with some commenting that only local government enabled them to put into practice all they had learned at University. Career development was also happening within posts, Township EHP 1 recently moving from a contract to a permanent position whilst three EHPs were promoted to Operations Manager posts shortly after the end of fieldwork.

When asked why they originally chose Urbington most EHPs cited better pay and conditions and/or a desire to be nearer home and family. Most had at least 2 years of Urbington experience, but even the least experienced (Suburban EHP 1 in Table 2.1) had worked as an EHP for two and a half years elsewhere. Further, there were no noticeable differences observed between the views and practices of younger and older EHPs that could be attributed to age or education, though there was some evidence that experience influenced regulation beyond the ten EHPs studied in depth. For example, it was recognised that inexperienced EHPs had, by definition, more limited knowledge of the law and its enforcement and this was sometimes used by managers to justify the need for close supervision. The case of a new food manufacturer in a formerly abandoned factory highlighted this well. Food production was stopped immediately following an inspection by Inner city EHP 5 and her Operations Manager, but a revisit by two inexperienced EHPs shortly afterwards gave the manufacturer the impression that he could restart immediately once some basic repairs were completed. This error was detected by the supervision process and corrected immediately but Managers remained shocked that their EHPs had made such decisions.

The working days of EHPs were largely determined by their position in the Departmental hierarchy in Figure 2.3. Central Unit Directors developed policy, coordinated the work of the regions and liaised with regional EHPs, politicians, the Director of Health, other departments and external organisations like Provincial and National governments. Regional EH Managers were mainly office based and focused on service delivery across their regions, liaising with Central Directors and politicians. Occasionally they worked in the field, particularly during non-routine and/or complex cases and when politicians were involved. Most Operations Managers had responsibility for four or five EHPs delivering District services but in each regional office, at least one Operations Manager was responsible for supervising officers, including EHPs, delivering Auxiliary services that included pest control, licensing and environmental pollution control. Operations Managers were mainly office based but frequently accompanied their EHPs/other officers in the field, particularly during complex cases. They also worked on complex cases, specialist projects and helped coordinate prosecutions.

Nine of the ten EHPs in Table 2.1 were District EHPs working as generalists primarily delivering local government environmental health services, known as municipal health services in South African law, in various roles that cut across the doughnut as summarised in Table 2.2. Health surveillance and food control took up most of their time but every day was different and District EHPs could find themselves working on cases across this great breadth of public health most days. A small number of Auxiliary EHPs, including Suburban EHP 1, specialized in environmental pollution control but they were few in number and all District EHPs were equipped with the knowledge and skills to work in this area as they did so most days.

Most daily work was in the field and face to face and this was something all valued, some adding that they did not want an office job. EHPs mostly worked alone and across areas determined by administrative boundaries (e.g. electoral wards), though sometimes other factors including resources and number of premises (e.g. a shopping mall) influenced the boundaries of the areas they covered. It was also common for EHPs to act up for their Operations Managers to cover periods when their manager was on leave or the post was vacant. Though their daily responsibilities and pay remained unchanged such opportunities provided valuable experience.

Table 2.2 The typical roles of Urbington EHPs delivering municipal health services

Municipal Health Service (Section 1 of the National Health Act, 2003*)*	*Examples of services observed*
Water quality monitoring	Microbiological sampling of river waters used for drinking and cooking in townships and informal settlements.
Food control	The surveillance and monitoring of the quality, safety and standards of foods on sale in streets and from food premises.
Waste management	Investigating the illegal dumping of solid wastes.
Health surveillance of premises	Monitoring and regulating health standards in food premises, other workplaces and housing.
Surveillance and prevention of communicable diseases	Investigating cases and outbreaks of infectious disease, monitoring standards of water, sanitation and hygiene.
Vector control	Monitoring rat activity in informal settlements, removing snakes from homes and areas.
Environmental pollution control	Investigating noise, air and water pollution complaints, regulating diesel vehicle emissions.
Disposal of the dead	Regulating the health standards of mortuaries.
Chemical safety	Monitoring and regulating operators and retailers involved in the manufacturing, application, transport and storage of chemicals.

Much of their work was proactive and unannounced, with EHPs generally flexible when encountering busy workplaces, though as Hutter (1988) found this was also dependent on whether they thought owners/managers were genuinely busy or being evasive. EHPs only booked proactive visits when issues like access or safety were problematic. Whether EHPs were accompanied during proactive visits was determined by the case itself (e.g. seriousness, staff availability) and their own preferences. Some preferred to work alone and report their findings afterwards, others gave people the option to accompany them which was usually accepted. All visits ended with meetings to review findings and the next steps. One manager commented that Urbington had lost a lot in the past by working reactively but by becoming more proactive once again environmental health was improving, particularly through improved surveillance and the targeting of hotspots. Indeed, proactive targets were now written into the performance agreements of EHPs as explored in Chapter 5.

For reactive work, complaints covered the breadth of environmental health but those related to waste management (e.g. illegal dumping), environmental pollution control (e.g. noise, smoke) and vector control were most common. Complaints reached EHPs via many sources including telephone calls, local politicians, visits to regional offices or during visits and meetings. The latter was particularly important in townships and informal settlements, with one EHP commenting:

> ... if you wait for them to come [with complaints] they are not going to come.
>
> (Township EHP 1)

Outside office hours complaints were recorded on answer machines and documented the next working day. Other sources of complaints included managers, politicians and other departments.

All EHPs had Urbington email accounts and computers but access to these was sometimes limited and EHPs usually corresponded with complainants by phone, letters and visits. All complaints were treated seriously by EHPs because, as Hutter (1988) found, their actions could be subject to greater public scrutiny than during proactive work, particularly when politicians were involved. The basic details of the complaint were recorded by the administrator/EHP and entered into a new electronic database, though the old regional office complaints book was also being maintained as a precaution. Where possible the GIS location of the complaint was pinpointed to gather further details (e.g. land ownership) and ensure allocation to the correct EHP. This worked well in theory, but in practice call operators lacked knowledge and were making mistakes like inputting the same details into the 'Address of Complaint' (i.e. where the potential offence happened) and 'Address of the Complainant' and therefore EHPs were extra cautious during visits until these facts were established.

Complaints were usually allocated to the EHP responsible for the area and, where possible, the EHP would try to witness the complaint, aware that delays could result in a loss of evidence. Responding to complaints quickly also helped EHPs maintain good relations with all parties, though investigations could be delayed by a lack of evidence, difficulties contacting complainants (e.g. no telephones) and cases involving other departments. Performance targets usually required

EHPs to initiate investigations within two working days of receipt, with a further three working days to initiate action. Reports for completed complaints were reviewed by managers before being sent to regional offices for archiving, though EHPs were also retaining hard copies given the fragility of the new database.

Suburban EHP 3 commented that the structure of the working day had changed little since he first started in the 1970s, though all EHPs now had their own cars. From Monday to Friday most were in their regional offices before 08:00 and busy with paperwork until around 10:00 when they started work in their areas. Around 15:00 they returned to their offices and spent the remaining time on administration before leaving around 16:30. EHPs normally didn't work evenings or weekends, instead answer phones recorded public complaints and EHPs would start responding the next working day. Non-routine work (e.g. noise monitoring of night clubs, blitz work targeting hot spots) and other activities (e.g. health promotion) sometimes required EHPs to work evenings or weekends and was popular when it was counted as overtime.

Paperwork was generally considered a necessary evil, one EHP summarising well an office culture that others described where:

> … if it's not written, then it's not done.
>
> (Township EHP 2)

Older EHPs recognised they now had far more discretion in their work than during apartheid but the quantities of paperwork had also increased as well as demands from managers and others. In each regional office, very small numbers of EH assistants and administrators supported EHPs in various ways (e.g. recording complaints, locating premises and owners) but EHPs completed their own paperwork and most preferred this to ensure timely completion and accuracy:

> … you cannot actually go outside there and then have another person write the report.
>
> (Inner city EHP 2)

The recording of daily, weekly and monthly performance data was a constant and each day all paperwork went to Operations Managers for review and approval. Achieving the right fieldwork-paperwork balance was a constant struggle amidst unpredictable workloads where, as one EHP put it, more problems could equal more paperwork. Managers recognised this and were generally supportive, but EHPs also valued how paperwork helped evidence their work to managers. EHPs also valued their discretion once in the field:

> ... you start at 8, leave at half past 4, you are in the office from 8 to 10, from 10 you should be out of the office because your work is at the field, come back 3 o'clock ... but [managers] rarely tell you exactly how to work ... you must just do it yourself.
>
> (Inner city EHP 3)

> ... you are given an area and then you supervise that area, whatever problems arise in that area, and then you are responsible. There is no one always behind you, 'where are you, what are you doing?' ... we are responsible people, that is very good about our job, there is not always somebody going with you ... you become responsible, you become adult, you know to supervise yourself.
>
> (Inner city EHP 2)

EHPs knew the limits to their discretion but all valued the flexibility it gave them in managing their working day, for example by organising their days to focus on areas or sectors with the greatest needs. Such decisions could create tensions with managers but EHPs maintained they were best placed to make these decisions and managers usually accepted this.

Note

1 Definitions of SMMEs vary by sector, turnover and asset value, but using numbers of full-time equivalent employees: medium enterprises employ < 200, small <50 and micro <5 (DTI, 2008).

References

Allison, M. (2002) Balancing responsibility for sanitation. *Social Science and Medicine*, Vol. 55 (9) 1539–1551.

ANC African National Congress (1994) *The Reconstruction and Development Programme: A Policy Framework*.

Andersson, N. and S. Marks (1988) Apartheid and health in the 1980s. *Social Science and Medicine*, Vol. 27 (7) 667–681.

Beall, J., O. Crankshaw and S. Parnell (2000) Local government, poverty reduction and inequality in Johannesburg. *Environment and Urbanisation*, Vol. 12 (1) 107–122.

Burke, S., I. Gray, K. Paterson and J. Meyrick (2002) *Environmental Health 2012: A Key Partner in Delivering the Public Health Agenda*. Health Development Agency, London, UK.

CEH Commission on Environmental Health (1997) *Agendas for Change*. Chadwick House Group Ltd., London, UK.

Chadwick, E. (1842/1965) *Report on the Sanitary Condition of the Labouring Population of Great Britain 1842 – edited by MW Flinn'*. Edinburgh University Press, Edinburgh, UK.

Coovadia, H., R. Jewkes, P. Barron, D. Sanders and D. McIntyre (2009) The health and health system of South Africa: historical roots of current public health challenges. *The Lancet*, Vol. 374 (9662) 817–834.

Cornell, S. (1996) Do environmental health officers practice public health? *Public Health* Vol. 110 73–75.

Dahgren, G. and M. Whitehead (1991) *Policies and Strategies to Promote Social Equity in Health*. Institute for Futures Studies, Stockholm, Sweden.

Dahlgren, G. and M. Whitehead (2021) The Dahlgren-Whitehead model of health determinants: 30 years on and still chasing rainbows. *Public Health*, Vol. 199 (October 2021) 20–24.

Davies, D., A. Crouch, F. Petersen and D. Williams-Wynn (1997) *Report of the First Phase of the Commission of Inquiry into Thor Chemicals' Terms of Reference*. Secretariat of the Commission of Inquiry, Pretoria, South Africa.

de Visser, J. (2005) *Developmental Local Government: A Case Study of South Africa*. Intersentia Publications, Antwerp, The Netherlands.

Dhesi, S. and Z. Lynch (2016) What next for environmental health? *Perspectives in Public Health* July 2016, Vol. 136 (4) 225–230.

DTI Department of Trade and Industry (2008) *Annual Review of Small Business in South Africa 2005–2007- Final Draft August 2008*. Department of Trade and Industry, Republic of South Africa.

Eales, K., Dau, S. and Phakati, N. (2002) Chapter 6: Environmental health. *South African Health Review 2002*. Health Systems Trust, Durban, South Africa.

Eyles, J. (1997) Environmental health research: setting an agenda by spinning our wheels or climbing a mountain? *Health and Place*, Vol. 3 (1) 1–13.

Ferrinho, P., P. Barron, E. Buch, J. Gear, A. Morris, F. Orkin, S. Bekker and A. Jeffrey (1991) Measuring environmental health status in Oukasie, 1987. *South African Medical Journal*, Vol. 79 29–31.

Hardoy, J., Mitlin, D. and Satterthwaite, D. (2001) *Environmental Problems in an Urbanising World: Finding Solutions for Cities in Africa, Asia and Latin America*. Earthscan Publications Ltd, London, UK.

Harpham, T., T. Lusty and P. Vaughan (eds.) (1988) *In the Shadow of the City: Community Health and the Urban Poor*. Oxford University Press, UK.

Huchzermeyer, M. (2004) From 'contravention of laws' to 'lack of rights': redefining the problem of informal settlements in South Africa, *Habitat International*, Vol. 28 333–347.

Hutter, B. (1988) *The Reasonable Arm of the Law? The Law Enforcement Procedures of Environmental Health Officers.* Clarendon Press, Oxford, UK.

Lewin, S., Urquhart, P., Strauss, N., Killian, D. and Hunt, C. (1998) *Linking Health and Environment in Cape Town, South Africa* Report produced by the Health Systems Division of the Medical Research Council of South Africa for the London School of Hygiene and Tropical Medicine.

Mabin, A. (1986) Labour, capital, class struggle and the origins of residential segregation in Kimberley, 1880–1920. *Journal of Historical Geography*, Vol. 12 (1) 4–26.

MacArthur, I. and Bonnefoy, X. (1997) *Environmental Health Services in Europe 1. An Overview of Practice in the 1990s* (WHO Regional Publications No. 76) WHO Regional Office for Europe, Copenhagen, Denmark.

Marmot, M. (2016) *The Health Gap: The Challenge of an Unequal World.* Bloomsbury, London, UK.

Mathee, A., Swanepoel, F. and Swart, A. (1999) *Chapter 20: Environmental Health Services, South African Health Review 1999.* Health Systems Trust, Durban, South Africa.

McDonald, D. (1997) Neither from above nor from below: municipal bureaucrats and environmental policy in Cape Town, South Africa Canadian. *Journal of African Studies*, Vol. 31 (2) 315–340.

McGranahan, G. (2007) International interview: Gordon McGranahan. *Environmental Health Practitioner Magazine* July 2007.

McGranahan, G. and D. Satterthwaite (2000) Environmental health or ecological sustainability? Reconciling the brown and green agendas in Urban Development in Pugh, C. (Ed.) *Sustainable Cities in Developing Countries: Theory and Practice at the Millennium.* Earthscan Publications Ltd., London, UK.

Mitlin, D. and D. Satterthwaite (2004) Chapter 1: Introduction, in Mitlin, D. and D. Satterthwaite (Eds.) *Empowering Squatter Citizen: Local Government, Civil Society and Urban Poverty Reduction.* Earthscan, London, UK.

National Treasury (2013) Chapter 6: Social security and the social wage in Republic of South Africa National Treasury. *National Budget Review 2013.* Communications Directorate, National Treasury, Pretoria, South Africa.

NDoH National Department of Health (2013) *National Environmental Health Policy* Government Gazette No. 37112 4 December 2013.

Norman R., D. Bradshaw, S. Lewin, S. Cairnross, N. Nannan and T. Vos (2010) Estimating the burden of disease attributable to four selected environmental risk factors in South Africa. *Reviews on Environmental Health*, Vol. 25 87–119.

Parnell, S. (1993) Creating racial privilege: the origins of South African Public Health and Town Planning Legislation. *Journal of Southern African Studies*, Vol. 19 (3) 471–488.

Parnell, S. (1991) Sanitation, segregation and the Natives (Urban Areas) Act: African exclusion from Johannesburg's Malay Location, 1897–1925. *Journal of Historical Geography*, Vol. 17 (3) 271–288.

Parnell, S. (2002) Winning the battles but losing the war: the racial segregation of Johannesburg under the Natives (Urban Areas) Act of 1923. *Journal of Historical Geography*, Vol. 28 (2) 258–281.

Petrie, J., Y. Burns and W. Bray (1992) Chapter 17: Air pollution, in R. Fuggle and M. Rabie (eds) *Environmental management in South Africa*. Juta & Co. Ltd, Cape Town, South Africa.

Proctor, M. (1986) Local and central state control of black settlement in Munsieville, Krugersdorp. *GeoJournal*, Vol. 12 (2) 167–172.

Rabie, M. and R. Fuggle (1992) Chapter 2: The rise of environmental concern, in R. Fuggle and M. Rabie (eds) *Environmental management in South Africa*, Juta & Co. Ltd, Cape Town, South Africa.

Raworth, K. (2017) A Doughnut for the Anthropocene: humanity's compass in the 21st century Lancet Planetary Health, Vol. 1 (2).

Rittel, H. and M. Webber (1973) Dilemmas in a general theory of planning, *Policy Sciences*, Vol. 4 155–169.

Rogerson, C. M. (1986) Feeding the common people of Johannesburg, 1930–1962. *Journal of Historical Geography*, Vol. 12 (1) 56–73.

Rosen, G. (1993) *A History of Public Health*. The Johns Hopkins University Press, Baltimore, USA.

RSA Republic of South Africa (1998) *White Paper on Local Government*. Department of Constitutional Development, Pretoria, South Africa.

SACN South African Cities Network (2004) *State of the Cities Report 2004*. South African Cities Network, Johannesburg, South Africa.

Sanders, D. and Chopra, M. (2006) Key challenges to achieving health for all in an inequitable society: the case of South Africa. *American Journal of Public Health*, Vol. 96 (1) 73–78.

Smith, K., C. Corvalan and T. Kjellstrom (1999) How much global Ill health is attributable to environmental factors? *Epidemiology* September 1999, Vol. 10 (5) 573–584.

SSA Statistics South Africa (2011a) *Census 2011 in Brief*. Statistics South Africa, Pretoria, South Africa.

SSA Statistics South Africa (2013) *Millennium Development Goals Country Report 2013*. Statistics South Africa, Pretoria, South Africa.

Steffen, K. Richardson, J. Rockström, S. Cornell, I. Fetzer, E. Bennett, R. Biggs, S. Carpenter, W. de Vries, C. de Wit, C. Folke, D. Gerten, J. Heinke, G. Mace, L. Persson, V. Ramanathan, B. Reyers and S. Sörlin (2015) Planetary boundaries: Guiding human development on a changing planet. *Science*, Vol. 347 (6223).

Steyn, P. (2008) Industry, pollution and the apartheid state in South Africa. *Scottish Association of Teachers of History, History Teaching Review Yearbook*, Vol. 22 67–75.

Steyn, P. and A. Wessels (2000) The emergence of new environmentalism in South Africa 1988–1992. *South African Historical Journal*, Vol. 42 210–231.

Sulla, V., P. Zikhali and P. F. Cuevas (2002) *Inequality in Southern Africa: An Assessment of the Southern African Customs Union (English)*. World Bank Group, Washington DC, USA.

Swanson, M. W. (1977) The sanitation syndrome: bubonic plague and urban native policy in the Cape Colony 1900–1909. *Journal of African History*, Vol. XVIII (3) 387–410.

Thomas, E., J. Seager and A. Mathee (2002) Environmental health challenges in South Africa: policy lessons from case studies. *Health & Place*, Vol. 8 251–261.

UN United Nations (1976) *The Vancouver Declaration on Human Settlements*. HABITAT United Nations Conference on Human Settlements, Vancouver, Canada, 31 May to 11 June 1976.

UNDP United Nations Development Programme (2015a) *Human Development Report 2015: Work for Human Development*. UNDP, New York, USA.

UNDP United Nations Development Programme (2015b) *Human Development Report 2015: Work for Human Development – Briefing Note for South Africa*. UNDP, New York, USA.

UNDP United Nations Development Programme (2003) *South Africa Human Development Report 2003, The Challenge of Sustainable development in South Africa: Unlocking People's Creativity*. Published for the United Nations Development Programme by Oxford University Press, Oxford, UK.

UNFPA United Nations Population Fund (2007) *State of World Population 2007: Unleashing the Potential of Urban Growth*. UNFPA, New York, USA.

WHO World Health Organisation (2008) *Our Cities, Our Health, Our Future: Acting on Social Determinants for Health Equity in Urban Settings*. Knowledge Network on Urban Settings, WHO Centre for Health Development, Kobe City, Japan.

3 How do EHPs tackle environmental health inequalities?

Introducing responsive regulation

My study was originally based on two theories of front-line public policy implementation I'd never encountered before during my own undergraduate environmental health training. Lipsky's classic theory of street-level bureaucracy draws on research from cities in the 1960s and 1970s America where poorly resourced officials, including EHPs, were being accused of distorting policies through their incompetence, bias, lack of cooperation with clients and resistance to change (Lipsky 1980). He critiqued these accounts by synthesising previous theories to describe street-level bureaucrats as public service workers with substantial discretion that gives them considerable power over service users and autonomy from their employers (Lipsky 1980). However, this power is accompanied by the problems of working in contexts not of their own choosing and in which resources are limited and demand far exceeds supply. Indeed, when faced with these dilemmas of public service:

> ... the decisions of street-level bureaucrats, the routines they establish, and the devices they invent to cope with uncertainties and work pressures, effectively become the public policies they carry out.
>
> (Lipsky 1980: xii)

More recently, Maynard-Moody and Musheno (2003) built on Lipsky's work in their own research on American public service workers. They identified Lipsky's bureaucrats as state agents but found they don't describe themselves as such. Instead, they liken themselves to citizen agents, making decisions based on their own identities as helpers or protectors and in response to individuals and

DOI: 10.1201/9781003009931-3

circumstances. Moralities are vital, with citizen agents more likely to help those considered worthy, whilst pragmatism is essential given the dilemmas of front-line work (Maynard-Moody and Musheno 2003). Sometimes state and citizen agent narratives exist in concert, for example when laws or policies support officials' views of people, but tensions are also inevitable such as when the law conflicts with officials' views of fairness or morality. Therefore street-level work can be considered:

> … as much a process of forming and enforcing identities – of both citizen-clients and street-level workers – as of delivering services and implementing policy.
>
> (Maynard-Moody and Musheno 2003:153)

A further breakthrough was the discovery of Hutter's (1988) research into UK local government EHPs which blended the core elements of the state and citizen agent within a broader socio-legal framework informed by theoretical perspectives from criminology, history, the law, organisational studies, political science and sociology. Other key studies of street-level environmental health regulators were identified from the UK (e.g. Hawkins 1984; Richardson et al. 1983) and Ghana (Crook and Ayee 2006). In the absence of research into South African EHPs other useful studies included Lewin et al.'s (1998) review of local government environmental health service reform in Cape Town, Beall et al.'s studies of local government reform in Johannesburg (2000a, 2000b and 2002) and the on-going work of the South African Cities Network (SACN 2004, 2006 and 2016).

Central to this developing framework was the concept of responsive regulation. Early descriptions of regulation echo Weber's (1947) and Foucault's (1977) descriptions of a 'bureaucratic surveillance state' responsible for deciding whether to punish or persuade offenders by what Hawkins (1984) calls compliance or deterrence/sanctioning strategies. Both seek to secure legal compliance and prevent offending behaviours but differ in how to achieve this. Compliance strategies seek to encourage compliance (Reiss 1984) by solving problems and preventing others from using methods like persuasion, negotiation and education (Hutter 1997). Here sanctions are primarily used as threats (Reiss 1984) over the long term and as a last resort when other options have failed (Hawkins 1984; Hutter 1988). In contrast, sanctioning strategies use deterrence to achieve compliance by detecting and punishing offenders to prevent future offending behaviours (Reiss 1984) or prohibit certain activities (Hutter 1999). Such strategies have been associated with the police and compliance strategies with regulators (Reiss 1984), including local government EHPs (Hutter 1988), though this distinction may be simplistic. For example, Hawkins reminds us that all law enforcers, whether police or regulators:

> will sometimes advise, instruct, exhort, bargain, or threaten, and will sometimes take formal steps to initiate the process of criminal prosecution (1991:428).

Others retain the compliance/deterrence distinction but identify different degrees of compliance-based approaches. For example, Hutter's (1988; 1997) work on UK EHPs and other regulators differentiated between so-called *persuasive strategies*, characterised by accommodation through education and negotiation over the long term and *insistent strategies*, also accommodating but less flexible and more likely to invoke formal action. More recent debate has combined compliance and deterrence strategies into a responsive model that instead questions when, not whether, regulators should punish or persuade. For example, Kagan and Scholz (1984) argue that regulators should respond to the motivation and ability of businesses to comply in accordance with different images of the regulations as summarised in Table 3.1.

Table 3.1 Kagan and Scholz's (1984) three images of the regulated

	Compliance motivations	*Response of regulator*
Amoral calculators	Profit-seeking	As the *police*, using less flexible, sanctioning strategies
Political citizens	Inclined towards compliance, but offending could result from a principled disagreement with laws considered authoritarian or unreasonable.	As a *politician*, trying to persuade businesses to act in the public interest but willing to compromise in response to legitimate business concerns or use sanctions if non-compliance continues.
Organizationally incompetent	Inclined towards compliance but offending results from organizational failures.	Working with regulated as both *consultant* and *educator*.

These three images are not mutually exclusive and Kagan and Scholz warn how they lack predictive power, for example, where a regulator risks exploitation by the bad apple whose perceived incompetence is actually an amoral calculation. They also stress responsiveness, with regulators:

> … prepared to shift from strict policemen, to politician, to consultant and back again according to their analysis of the particular case.
>
> (Kagan and Scholz 1984:86)

These images were further developed by Ayres and Braithwaite (1992) who combined compliance and deterrence to create responsive regulation as summarised in the pyramid on the right of Figure 3.2, adapted here for Urbington's EHPs. The least intervention occurs at its base using persuasion in the first instance, but the regulator escalates their interventions upwards on a tit-for-tat basis using increasingly punitive approaches to secure compliance. Likewise, regulators can de-escalate once compliance is achieved and sustained. In this way, Braithwaite (2007) argues, regulators can secure a moral commitment to the law that rewards compliance, deters non-compliance and enables potential offenders to change their ways. The aim is to provide a flexible and measured framework for regulation that is responsive to the changing behaviours and motivations of the regulated and other factors explored over the following chapters. But before exploring responsive regulation in more detail I now describe the two pathways Urbington EHPs use to regulate environmental health.

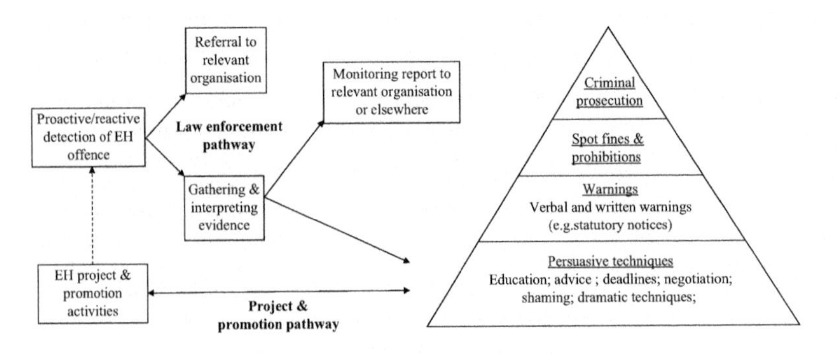

Figure 3.1 The stages of regulation by Urbington EHPs.

Law enforcement pathway

This traditional pathway began with the detection of potential environmental health offending by proactive or reactive means. The word 'potential' is important because at this stage it could be hard to establish whether an offence has been committed without further investigation. Most proactive work was unannounced but before accessing any premises/areas EHPs would show their identification (ID) and introduce themselves as '[name] from the Health Department' and explain the purpose of their visit. Depending on the case, they would then ask to speak to the owner/person in charge and discuss their visit in more detail before starting any investigations. If EHPs knew the premises these introductions were brief and did not involve ID, but when visiting for the first time EHPs were careful to clarify the purpose of their visit for many reasons, including a wish to build cooperative working relations.

Matters obviously not relevant to EHPs were referred to the correct department or external regulator. All other matters were then subject to further investigation by EHPs using methods including observation, interviews and the analysis of documents. Back in the office, other checks might include the past history of the business, previous complaints and/or background checks of GIS records. Other organisations might also be contacted and together this evidence is informed when referral to another organisation is necessary or if EHPs have complete or shared responsibilities for a case. Responsibility could be shared because EH regulation is fragmented in the municipality itself and between provincial and national governments, as set out in Table 4.2.

In undertaking monitoring roles for other organisations and provincial government, one manager likened EHPs to the 'eyes and ears' of Urbington. EHPs considered the ability to work with so many organisations as a key skill and monitoring reports were drafted by them and then reviewed by managers before being referred to the right organisation(s). In theory, this ended the EHP's involvement, but in practice monitoring usually continued because conditions changed or problems continued.

Moving to the pyramid itself its shape was confirmed by three sources of evidence. First, observation found that Urbington EHPs' routine work was dominated by persuasive approaches, particularly education and advice. Written warnings (e.g. statutory notices) were issued after some visits but only one spot fine was served during 89 observed routine visits. Second, during fieldwork law enforcement trend data were unavailable but the pyramid mirrored the performance targets set for EHPs and suggested a probability of around one spot fine served per 50 routine visits, assuming that EHPs at least meet their annual targets. Third, law enforcement data from one campaign targeting known offending hotspots suggested a far steeper pyramid (e.g. one spot fine per 7 visits), but even then activities lower down the pyramid were still predominant.

The stage at which cases entered the regulatory pyramid and the speed they moved up/down depended on the case and the factors explored across Chapters 4 to 7 but EHPs used mainly persuasive approaches towards achieving compliance with the law. If these initial attempts were unsuccessful they would be followed by negotiations where EHPs typically told stories or recalled events about similar cases to appeal to a person's sense of responsibility to their customers or the public. The agreement of deadlines was useful here because, as Hutter (1988) found, it suggested reasonableness on the part of the EHP whilst obliging the offender to carry out the work needed in the agreed time period. Shaming businesses into compliance, for example by comparison with better premises nearby, was another technique used. Attempts to bluff were also important and included making demands or threats (e.g. 'you know I could close you down', 'I could fine you R500') that might be difficult to enforce if called. Dramatic techniques sometimes helped EHPs reinforce their demands and included raising voices or asking people to point out offences themselves.

Failure to secure compliance would usually result in escalation to the next level of the pyramid, with verbal and written warnings. The former included threats of further escalation if offending continued, particularly the serving of a spot fine. Threats could also involve managers, other organisations or the courts themselves, tactics that helped shift responsibility for action onto the offender in a reasonable way. The most common written warnings were statutory notices but they also included reports or letters, some based on standard formats and checklists, though EHPs valued the discretion to write their own letters. Details like GIS map coordinates were also included and digital photos were increasingly being used to give reports more impact, particularly when other organisations or politicians were involved. Written warnings typically ended with a standard sentence like 'your co-operation will be greatly appreciated, but failure to comply may result in legal proceedings being taken against you'. Some warnings also cautioned that a spot fine could result from continued non-compliance and stated the maximum fine for that offence upon conviction. In the field statutory notices were written on blank and indexed carbon copy sheets, with the top sheet given to the offender at the end of the visit. Of the two remaining copies, one was retained and filed by the EHP and the other was sent to their Operations Manager for review. Hand-written warnings were usually typed up by EHPs, reviewed by their manager and then posted to the offender where possible (e.g. some informal traders/settlements lacked postal addresses), or hand delivered and filed for future reference.

Continued non-compliance could result in another written warning or a spot fine that incorporates both a fine and a court summons using Urbington Police Department's standard carbon copy form for road traffic offenders. EHPs completed this form on site with details of the offence and the next regional court date (EHPs carried lists of local court dates in their bags) should the offender want to contest the offence. Further information on how to pay the fine or appeal was provided on the form in Afrikaans and English, including a warning that the failure to comply could result in arrest and sentencing to a fine not exceeding R1500 (~$210 in 2007) or imprisonment for a period not exceeding 3 months. For those admitting guilt, details of the various payment methods of the Department of Transportation are included. Those disputing the offence should inform the public prosecutor in writing before the payment of the fine deadline and were recommended to use a lawyer or apply for legal aid via the court. The original copy of the spot fine was left with the offender, another was filed by the EHP and the third was submitted to the local duty Sergeant/Inspector of Urbington Police for review and, in theory, entry into the criminal justice system database.

In Figure 3.1, prohibitions included the closure of premises, the suspension/withdrawal of licenses/permits and other procedures like court orders to take control of premises/buildings or confiscate equipment (e.g. music amplifiers). These were not observed and only a few EHPs had the experience but senior managers were busy drafting guidance for EHPs who were keen to use them more because of the problems with the spot fine explored in Chapter 6. At the pyramid's peak, criminal prosecution was the highest sanction and the nature of the spot fine meant that, in theory, prosecution happened automatically if an offender chose to contest their fine in court. Offenders not paying their fines and then failing to appear in court were, in theory, automatically issued warrants of arrest. EHPs could also force such cases to court by requesting that public prosecutors serve a *Notice to Appear in Court* on the offender. During fieldwork, Operations Managers were preparing prosecution cases but the involvement of the EHPs observed was limited.

Project and promotion pathway

Returning to Figure 3.1, most EHPs were also engaged in a second pathway comprising ongoing EH projects and promotional activities with schools, businesses, community groups and non-government organisations (NGOs) amongst others. These interventions are separate from the law enforcement pathway because they often do not originate from offending and were initiated by EHPs or local actors themselves. A broken arrow is used to link the pathways because sometimes these activities led to the detection of offending and they were often shaped by the law itself.

Some EHPs provided environmental health courses in their regions on many topics including food hygiene training for informal food handlers or Lead (the harmful metal) awareness training for pre-schools. Each course was structured around the technical and practical aspects of the subject and tailored to the requirements of delegates, particularly those from the informal sector, but EHPs constantly reminded them of their legal responsibilities and warned that non-compliance could result in punishment.

Fieldwork coincided with a city festival featuring stands from many Urbington services, including Environmental Health, with talks by EHPs to visiting school children about their work and topics including waste management and climate change. Global environmental health days were celebrated each year and EHPs were working with schools to plan activities for 'World No Tobacco' and 'World Environment' days, including healthy environment poster competitions and community environmental workshops. EHPs were also engaged in activities to raise public awareness of their services and environmental health more generally which included the distribution of advisory leaflets in residential areas, shopping malls and workplaces that explained their services, gave instructions on how to complain and details of local EHPs. Urbington's website also published this information alongside links to priority areas including permits and licenses (including application forms), illegal dumping and guidance on food hygiene.

EHPs worked with community groups in different ways. Attending meetings gave opportunities to provide information and education on environmental health and other Urbington services and receive complaints and intelligence on local issues that might otherwise remain undetected. Further, EHPs could also make referrals to other services where appropriate. One EHP had recently spoken at a local youth event about illegal dumping and law enforcement and was planning to run more events like this in the future:

> [By educating children] you make your programme sustainable, because they are the future leaders and they can EASILY take the information back to their homes.
>
> (Inner city EHP 4)

Work with charities across Urbington also included the provision and maintenance of basic services for vulnerable groups like refugees and those affected by HIV/AIDS and EHPs recognised that maintaining good relations with them could be mutually beneficial:

> They know us, that we are there, whenever we need their assistance we call them on board, or maybe their chairman and sit down and discuss ... we are working hand in hand, you know it's good to work like that, that can make our job easier at the end of the day because they are actually there every day, they are staying there, they know the situation.
>
> (Inner city EHP 2)

As well as contributing to local projects some EHPs had also developed their own in response to problems and their wider interests. They included an ongoing clean-up campaign targeting ten hostels and working with residents, Ward Councillors and other services (e.g. waste management) to clean the hostels and surrounding areas and provide workshops on healthy lifestyles and the municipal By-laws. Halfway through the campaign more than 400 residents, including women, children and the unemployed had participated. A group of inner-city EHPs had developed a smokeless coal stove and, as part of a wider programme to improve air quality and empower small businesses, had donated some to informal township traders and were planning similar future donations. Suburban EHP 2 was a former gardener and was working with township residents to create a park.

Persistent illegal dumping problems, sometimes of a scale that could block roads, had prompted EHPs to organise local projects to complement their law enforcement work. One EHP reduced illegal dumping by involving local people in the monitoring of known hot-spots. They now did not hesitate to call him if they saw anything and other EHPs were replicating this in their areas. Another EHP had started a similar project near his own home, partly by shaming (in his own words) his neighbours into action by asking what they had done to make a difference to their local environment beyond just complaining to Urbington! Indeed, EHPs were constantly trying to persuade people to take more responsibility for their environmental health, one municipal briefing began:

> Urbington municipality takes the protection, conservation and enhancement of its environment seriously. However, inconsiderate people litter our streets, dump refuse and waste on vacant stands and open spaces and pollute our underground water resources, soil, streams and air. The monitoring and control of illegal activities, pollution and dumping cost Urbington a great amount of money. This cannot be allowed, as these wasted funds are needed for urgent projects such as housing, infrastructure and other services. Urbington municipality cannot protect, conserve and enhance the environment on its own and therefore requests the assistance and commitment of all stakeholders in combating these crimes.

The briefing then requested the public to inform EHPs of anyone who commits any offences listed there, confidentiality guaranteed and warns that zero tolerance will be applied with stiff fines for those found guilty by the courts. By prosecuting offenders, Urbington argued, they were protecting the rights of South Africans.

Environmental health regulation as governance

Lastly, my description of how Urbington EHPs regulate environmental health ends by embedding the responsive regulation pyramid into a wider framework of governance that can be explained by a brief return to two theories of modern state power used by Crook (2007) to explain why local government EHPs do what they do. The first draws on Weber's bureaucracy (1947) and MacDonagh's (1958) work by associating local government EHPs, or sanitary inspectors/inspectors of a nuisance as they were known in the 19th century, with the gradual movement away from a laissez-faire state in the late Victorian period towards an increasingly bureaucratic and interventionist state characterised here by:

- The gradual emergence of professionally qualified EHPs
- Appointed by local government
- Bound by rules and working in hierarchies
- Exercising considerable discretion to carry out their legal duties.

The second theory combines Weber's bureaucracy (1947) with Foucault's (1977) work by associating regulation with the rise of the bureaucratic surveillance state in which the professional local government EHP is instead embedded within a theory of social power characterised by a belief in the legality of rules and the rights of rule-bound EHPs to issue environmental health commands to discipline and control populations. Historical accounts of colonial and apartheid South Africa (e.g. Andersson and Marks 1988; Parnell, 1993; Rogerson, 1986; Swanson, 1977) depict this, particularly the state's intervention on environmental health grounds to justify the racial segregation, surveillance and control of the urban space to safeguard the public health of the White minority. But Crook (2007) also argues that these top-down theories obscure the personal nature of regulation that comes to life across this book and its operation within a critical and sometimes hostile public sphere. Therefore, he argues, regulation by local government EHPs is better understood as a form of surveillance and governance where:

> ... [p]ower circulates between and inhabits all these agents [state and society, experts and public] as they, by turns, resist and co-operate with one another ... in this way freedom is not a goal but a means of liberal governance, a process it works through as a form, however messy, of social ordering ... governance *was* the struggles inspectors endured and sought to overcome, which informed all aspects of their job, from direct encounters with the public to the ongoing battle for greater professional independence.
>
> (Crook 2007:393)

Returning to South Africa, though 'freedom' in the form of democracy wasn't achieved until the 1990s evidence of similar struggles includes attempts by colonial local authorities to enact segregationist laws on environmental health grounds that were sometimes thwarted by local authorities reluctant to finance alternative housing and services for Africans (Parnell 1991), resistance from African property owners (Swanson 1977) and from landlords and businesses who wanted Africans to live closer to work (Parnell 2002). From the late 1970s, state surveillance and control were met by the increasing resistance of oppressed groups, sanctions on the economy and the demands of capital for a more skilled and stable urban workforce that gradually weakened the state's power (Andersson and Marks 1988). This period also prompted the gradual development of new environmental health infrastructure and services for non-Whites, particularly in rural and self-governing 'homeland' areas though these improvements were fragmented and underfunded (Agenbag and Balfour-Kaipa 2008; Andersson and Marks 1988). From the 1990s the developmental vision for South Africa's local government EHPs outlined in Chapter 4 seems also seems far removed from that of a bureaucratic surveillance state, more recent studies (e.g. Barnes 2007; Lewin et al. 1998) suggesting the public remains critical, and sometimes hostile, to their work.

A model of governance is used to capture this complexity, with governance defined as the 'patterns of interaction between civil society and government' (Allison 2002: 1540). Here government focuses on Urbington's EHPs but also includes other municipal services, provincial and national government and the criminal justice system. Civil society constitutes those individuals and institutions outside government control (Harpham and Boateng 1997) including businesses and the public, charities, the media and professional organisations.

The governance model in Figure 3.2, therefore, embeds Figure 3.1 within a wider framework to illustrate how the decisions of EHPs are constantly shaped by individual, legal and organisational factors as well as their relations with the regulated and the context in which they work as explored over the coming chapters. Circular shapes are used here to depict the blurred boundaries of governance where these factors frequently overlap and power circulates between various actors. A funnel metaphor adapted from Kaufman (1960/2006) works well here, where regulation by local government EHP looks like a vast funnel with the EHP at its throat: all the factors above pour out materials which, mixed and blended by the EHP, emerge in a stream of regulatory action in the field.

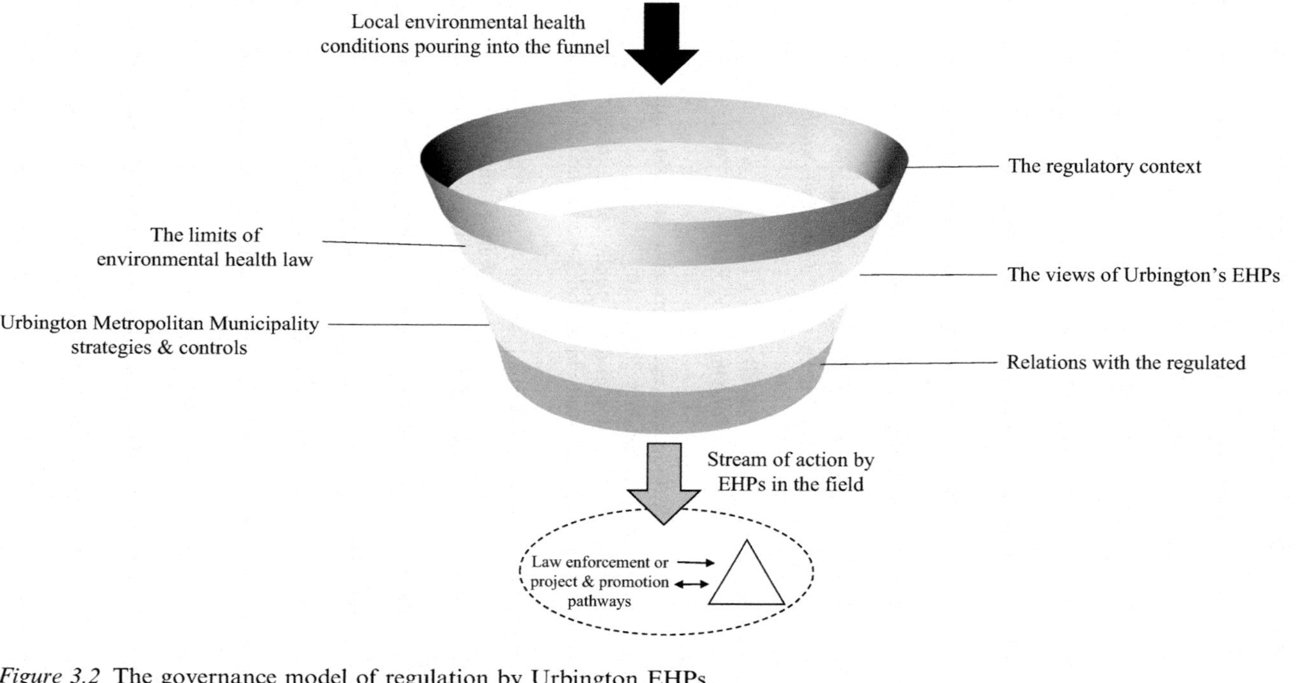

Figure 3.2 The governance model of regulation by Urbington EHPs.

References

Agenbag, M. and T. Balfour-Kaipa (2008) Chapter 10: Developments in environmental health, in Barron P. and J. Roma-Reardon (Eds) *South African Health Review 2008*. Health Systems Trust, Durban, South Africa.

Allison, M. (2002) Balancing responsibility for sanitation. *Social Science and Medicine*. Vol. 55 (9) 1539–1551.

Andersson, N. and S. Marks (1988) Apartheid and health in the 1980s. *Social Science and Medicine*. Vol. 27 (7) 667–681.

Ayres, I. and J. Braithwaite, (1992) *Responsive Regulation*. Oxford University Press, Oxford, UK.

Barnes, B. (2007) The politics of behaviour change for environmental health promotion in developing countries. *Journal of Health Psychology*, Vol. 12 (3) 531–538.

Beall, J., O. Crankshaw and S. Parnell (2000a) Local government, poverty reduction and inequality in Johannesburg. *Environment and Urbanisation*, Vol. 12 (1) 107–122.

Beall, J., O. Crankshaw and S. Parnell (2000b) Victims, villains and fixers: the urban environment and Johannesburg's poor. *Journal of Southern African Studies*, Vol. 26 (4) 833–855.

Beall, J., O. Crankshaw and S. Parnell (2002) *Uniting a divided city: governance and social exclusion in Johannesburg*. Earthscan Publications Ltd, London, UK.

Braithwaite, V. (2007) Responsive regulation and taxation: introduction. *Law and Policy*, Vol. 29 (1) January 2007 3–10.

Crook, T. (2007) Sanitary inspection and the public sphere in late Victorian and Edwardian Britain: a case study in liberal governance. *Social History*, Vol. 32 (4) 369–393.

Crook, R. and J. Ayee (2006) Urban service partnerships, 'street-level bureaucrats' and environmental sanitation in Kumasi and Accra, Ghana: coping with organisational change in the public bureaucracy. *Development Policy Review*, Vol. 24 (1) 51–73.

Foucault, M. (1977) *Discipline and Punish: The Birth of the Prison*. Penguin Books, London, UK.

Harpham, T. and K. Boateng (1997) Urban governance in relation to the operation of urban services in developing countries. *Habitat International*, Vol. 21 (1) 65–77.

Hawkins, K. (1991) Enforcing regulation: more of the same from Pearce and Tombs. *British Journal of Criminology*, Vol. 31 (4) 427–430.

Hawkins, K. (1984) *Environment and Enforcement: Regulation and the Social Definition of Pollution*. Clarendon Press, Oxford, UK.

Hutter, B. (1997) *Compliance: Regulation and Environment*. Clarendon Press, Oxford, UK.

Hutter, B. (1999) Socio-legal perspectives on environmental law: an overview, in B. Hutter, (ed.) *A Reader in Environmental Law*. Oxford University Press, UK.

Hutter, B. (1988) *The Reasonable Arm of the Law? The Law Enforcement Procedures of Environmental Health Officers*. Clarendon Press, Oxford, UK.

Kagan, R. and J. Scholz (1984) The 'criminology of the corporation' and regulatory enforcement strategies, in K. Hawkins and J. Thomas (Eds.) *Enforcing Regulation*. Kluwer-Nijhoff, Boston, USA.

Kaufman, H. (1960/2006) *The Forest Ranger: a Study in Administrative Behaviour*. Resources for the future, Washington, USA.

Lewin, S., P. Urquhart, N. Strauss, Killian, D. and C. Hunt (1998) *Linking Health and Environment in Cape Town, South Africa* Report produced by the Health Systems Division of the Medical Research Council of South Africa for the London School of Hygiene and Tropical Medicine.

Lipsky, M. (1980) *Street-level Bureaucracy: Dilemmas of the Individual in Public Services*. Russell Sage Foundation, New York, USA.

MacDonagh, O. (1958) The nineteenth century revolution in government: a reappraisal. *The Historical Journal*, Vol. 1 (1) 52–67.

Maynard-Moody, S. and Musheno, M. (2003) *Cops, teachers, counsellors: stories from the front lines of public services* University of Michigan Press, USA.

Parnell, S. (1993) Creating racial privilege: the origins of South African Public health and town planning legislation. *Journal of Southern African Studies*, Vol. 19 (3) 471–488.

Parnell, S. (1991) Sanitation, segregation and the Natives (Urban Areas) Act: African exclusion from Johannesburg's Malay Location, 1897–1925. *Journal of Historical Geography*, Vol. 17 (3) 271–288.

Parnell, S. (2002) Winning the battles but losing the war: the racial segregation of Johannesburg under the Natives (Urban Areas) Act of 1923. *Journal of Historical Geography*, Vol. 28 (2) 258–281.

Reiss, A. (1984) Selecting strategies of social control over organisational life, in K. Hawkins and J. Thomas (Eds.) *Enforcing Regulation*. Kluwer-Nijhoff, Boston, USA.

Richardson, G., A. Ogus and P. Burrows (1983) *Policing Pollution: A Study of Regulation and Enforcement*. Oxford University Press, Oxford, UK.

Rogerson, C. M. (1986) Feeding the common people of Johannesburg, 1930–1962. *Journal of Historical Geography*, Vol. 12 (1) 56–73.

SACN South African Cities Network (2004) *State of the Cities Report 2004*. South African Cities Network, Johannesburg, South Africa.

SACN South African Cities Network (2006) *State of the Cities Report 2006*. South African Cities Network, Johannesburg, South Africa.

SACN South African Cities Network (2016) *State of the Cities Report 2016*. South African Cities Network, Johannesburg, South Africa.

Swanson, M. W. (1977) The sanitation syndrome: bubonic plague and urban native policy in the Cape Colony 1900–1909. *Journal of African History*, Vol. XVIII (3) 387–410.

Weber, M. (1947) *The Theory of Social and Economic Organisation*. The Free Press, New York, USA.

4 The regulatory context and limits of environmental health law

This chapter begins by exploring the outer layer of the governance model, Urbington's regulatory context, by revisiting the historical recollections of some of its EHPs before briefly examining the influence of national macroeconomic policies and public and media attitudes towards environmental health. It then reviews the vast legal framework for environmental health that comes without a guidebook and why its many uncertainties frequently favour persuasion over punishment.

The regulatory context

Five EHPs had worked in Urbington since the 1970s–1980s and confirmed the predominance of law enforcement and how their work was inherently unfair to non-White people and businesses. Suburban EHP 3 confirmed this too but volunteered a more nuanced account. He first worked as a township EHP in the 1970s with responsibilities for measles and polio immunisation and supervising bucket latrines before being given more responsibility for public health by law enforcement across the township and eventually all Urbington. Back then national government was deliberately selecting, training and then posting inspectors to areas with the worst environmental health. Those appointed by the self-governing rural homeland governments were not very effective because they were so few in number (e.g. 1 or 2 per homeland) and had to travel long distances. By contrast, urban environmental health was much better in Suburban EHP 3's view due to population controls (e.g. the Pass laws) and the greater resources available. He typically inspected around 50 homes, hostels and businesses per day and often worked in areas where he did not initially speak the language but soon learned it. Threats from

DOI: 10.1201/9781003009931-4

township gangsters and the public were common, the latter often commenting:

> How can you, a fellow African?
>
> (Suburban EHP 3 imitating member of public)

In response Suburban EHP 3 thought it important to have a job, he was proud of his work and gradually progressed beyond the townships by inspecting so-called Grey food businesses, typically run by Greeks or Portuguese and serving factory workers. Initially, his White managers would revisit the premises to communicate Suburban EHP 3's findings to the White owners/managers but he was eventually authorised to do this himself. Fines and prosecutions for offending were common and he spent a long time in the courts which he didn't think was fair, but it sometimes helped improve local conditions. When prosecuting White people Suburban EHP 3's presence would often be met with outrage by the defendants:

> How dare a Black man tell us what to do!
>
> Who is this Black monkey?
>
> (Suburban EHP 3 imitating defendants in court)

But Suburban EHP 3 usually won his cases and magistrates sometimes held defendants in contempt of court for such comments. These historical accounts confirmed the role of EHPs in maintaining environmental health inequalities during apartheid that today's EHPs understandably wanted to distance themselves from. However, Suburban EHP 3's recollections suggest more complex realities in which resistance by the public and businesses was ever present, occasional opportunities to improve environmental health existed and Black African EHPs could get promoted.

Local government EHPs also find their work shaped by global macroeconomic policy. In the UK a powerful and sustained deregulatory rhetoric has characterised neoliberal macroeconomic policy since the 1980s, with regulators including local government EHPs accused of inefficiency and burdening businesses with red tape (Hutter 2005; Tombs 2016). Some liken this to an ideological war on law enforcement by the government (James et al. 2013) with inspectors encouraged to think more like social workers than regulators because the prosecution may not be 'the right thing to do' (Hawkins 2002:118). The South African government's shift towards more neoliberal macroeconomic policies is widely recognised (Coovadia et al. 2009; Habib and Padayachee 2000) and characterised by the move away from a state-led, integrated and developmental framework with environmental health at the fore (e.g. ANC 1994), towards decentralisation and more strategic government intervention alongside reduced state spending, an accelerated program of privatization, deregulation and trade and industry liberalization (e.g. RSA 1996). To quote from one influential neoliberal text of this time the role of the state changed to 'steering, not rowing' (Osborne and Gaebler 1992) and whilst I found no direct evidence of neoliberal pressures on Urbington EHPs to adopt more persuasive approaches, other neoliberal factors continuously influenced their work.

The structure of Urbington and the performance of its workforce were shaped by neoliberal policy. For example, EHPs were considered contractors to their Department of Health client whilst, in accordance with Hood's (1991) new public management approaches, their work was inherently results-oriented and shaped by performance targets focused on monthly outputs (e.g. numbers of inspections, complaints, notices served) that broadly conformed to the regulatory pyramid. Further, whilst there was no mention of plans for the future privatization or outsourcing of EHPs during fieldwork similar contractor-client structures are well established across South African local government and such arrangements are established in influential countries like the UK (Tombs 2016).

Lastly, EHPs were very conscious of the public and media attitudes to their work and the tensions caused by the huge inequalities in their areas. At the one extreme were affluent suburbs with the highest standards and most vocal residents:

> If you happen to get one [a rich complainant] they tell you how rich they are, how they can talk to their lawyer and you don't have to communicate directly with them. So if you are not familiar with the Act and the Regulation they can shut you out … so when you come they don't look at you as a professional, they just measure you concerning their wealth. It seems like you are too poor to come talk to them, 'send someone else' they said to me!
>
> (Township EHP 1)

On the other were areas with the worst environmental health and the least vocal residents to whom EHPs responded in many different ways, including greater monitoring and more education and awareness-raising activities but this was a persistent challenge. EHPs often observed general improvements in their areas but feared these were not sustainable. For example, Township EHP 2 believed there was now much less serious crime (e.g. car-jacking, rape) in some of her areas, but it remained a problem in others. Other EHPs were concerned that drugs and prostitution were feeding wider problems like overcrowding that spread into neighbouring areas. Migration to the city was also creating many challenges and opportunities for exploitation:

> Now the landlords of the people that stayed in these big houses ... realised that whilst there's money coming in they started subdividing the houses, getting people in, hence we have these overcrowding problems. Because now these houses were designed for families and we have forty people staying in a house designed for one family, and what happens to the infrastructure, blockages, problems of refuse removals, apart from the overcrowdings itself? Then also the neglect of the properties needing to be maintained ... So problems become MUCH BIGGER because of that.
>
> (Suburban EHP 3)

Indeed people moved in and out of areas, homes and businesses constantly, EHPs commenting that they were never sure what they would find, some likening their work to constantly 'putting out fires'.

Though EHPs thought their relations with the public were improving everything was still new, including these new approaches to working and it would take time for all to adjust:

> Our communities were used to apartheid for some time. This time the legislation was very prescriptive, Blacks have ... Whites have ... Everything came from the Government. Now there are few complaints from the community and yet conditions remain unsatisfactory, even appalling.
>
> (Township Operations Manager)

> ... I usually tell them, compared to apartheid while you were born for like 60 years under oppression, suddenly you are free but I believe your mind is still born dead because there are so many things that you got used to doing a certain way. Things have changed on paper ... but they actually come slowly. So I'm thinking people got used to staying in dirty environments, there were no By-Laws, anybody could do whatever they want, whatever. So now you come in, the Council, and you're telling

> them to paint their shop, you're telling them to put in hand wash basins, nail brush, soap and they are like 'Aha, you must be kidding me!'
>
> (Township EHP 2)

Encounters with a distrusting and occasionally hostile public were common across all Urbington. Those in more affluent suburbs expected high levels of services, were willing to complain and had little faith in EHPs to sort out their problems. In poorer areas, bad environmental health conditions made things difficult from the start, whilst fears that EHPs might close a business or evict someone could make matters worse. EHPs recognised the utility of the media for communicating their work and often asked them to cover projects or promotional activities. But EHPs commented that the media were rarely interested unless there were serious, visible problems, when their requests could backfire:

> Let's say there are blockages in the squatter camp, then [the media] will go there to shoot, to put it in the first page, telling the people that the Councillor and the EHPs are not doing their work … maybe the blocked drain happens on Friday late then I need to come here on Monday, they will say … 'the whole week it was like this and the health inspectors did nothing, and the councillors did nothing'.
>
> (Suburban EHP 2)

> … what happens is the newspapers and everybody will say 'the environmental health of people in [my area] is put at risk. Who exposed them? It is the EHP of the area'
>
> (Inner city EHP 1)

During fieldwork articles on Urbington's environmental health appeared regularly, particularly when highlighted by senior politicians and policymakers, but the focus was often on how local actors including Urbington, its EHPs and the courts sometimes made matters worse. For example following one recent Mayoral tour one Inner City EHP was furious with the Mayor's descriptions in the media of the 'appalling conditions' in his areas and accused the Mayor of knowing nothing about his work there but he was also expecting a hard time as a result. Local newspaper coverage was directed towards more serious issues that Suburban EHP 2 suspected were driven largely by the need to sell newspapers, not bearing witness or holding individuals and organisations to account. However, Urbington EHPs were clearly engaged because their office walls were covered in newspaper articles about their areas, particularly housing cases where it was not uncommon for 'environmental health inspectors' to be blamed for poor conditions or evictions.

Rights and reasonableness: the *Constitutional* framework for environmental health

The *Constitution* defines the environmental health rights of South Africans, the organisational structures and procedures for regulation and has been the highest law in South Africa following its publication in December 1996 after nearly 2 years of public consultation. The *Constitution* created local government across the whole of South Africa, including metropolitan municipalities like Urbington, and recognises local government as a distinctive sphere of government but one that is simultaneously equal, interdependent and interrelated to the national and provincial spheres of government. These relationships are enshrined in the principles of cooperative government and include a duty on all spheres of government to secure the well-being of the people and for a local government committed to:

- provide democratic and accountable government for local communities;
- ensure the provision of services to communities in a sustainable manner;
- promote social and economic development;
- promote a safe and healthy environment; and
- encourage the involvement of communities and community organisations in the matters of local government

(Section 152. (1))

In 1998 this mandate was extended by the *White Paper on Local Government* and its vision for a developmental local government:

> ... committed to working with citizens and groups within the community to find sustainable ways to meet their social, economic and material needs and improve the quality of their lives
>
> (RSA, 1998:23)

The *Constitution's* Bill of Rights also sets out the environmental health rights of South Africans, particularly the fundamental right 'to an environment that is not harmful to their health or well-being (Section 24a) that operates alongside the socioeconomic right:

> '... to have the environment protected, for the benefit of present and future generations, through reasonable legislative and other measures that:
>
> i prevent pollution and ecological degradation;
> ii promote conservation; and
> iii secure ecologically sustainable development and use of natural resources while promoting justifiable economic and social development
>
> (Section 24b)

Other socioeconomic environmental health rights include the right to access adequate housing and to health care services, sufficient food and water and social security alongside rights for every child to these basic needs. The *Constitution* also shapes how national, provincial and local government should deliver these rights by including the fundamental rights of everyone to:

- equality before the law and equal protection and benefit of the law (Section 9 (1));
- dignity and to have this respected and protected (Section 10); and
- administrative action that is lawful, reasonable and procedurally fair (Section 33 (1)).

These rights are further enacted across the law, including taking measures to prevent corruption in administration (*Local Government: Municipal Systems Act, 2000*) and conferring on 'health officers' various powers to investigate potential offending including the ability to enter premises at any reasonable time (*National Health Act, 2003*). Reasonableness is also written across Urbington's own *Public Health By-laws*, for example:

> In dealing with matters affecting public health the Council must -
>
> a Adopt a cautious and risk-averse approach;
> b Prioritise the collective interests of the people of the municipal area, and of South Africa, over the interests of any specific interest group or sector of society;

c Take account of historic inequalities in the management and regulation of activities that may have an adverse impact on public health and redress these inequalities in an equitable and non-discriminatory manner;
d Adopt a long-term perspective that takes account of the interests of future generations; and
e Take account of, and wherever possible without compromising public health, minimise any adverse effects on other living organisms and ecosystems

(Extract from *Urbington Public Health By-laws*)

Urbington EHPs, therefore, have a wide-ranging *Constitutional* mandate and a duty to act reasonably set out across vast amounts of legislation but there was no roadmap to guide EHPs on how to regulate but national policy sometimes encouraged them towards more persuasive approaches.

Notable was the 1997 *White Paper on Transforming Public Service Delivery* whose eight Batho Pele ('People First' in Sotho-Tswana) principles for delivering public services included consultation with citizens, equal access to services, treating citizens with courtesy, providing information about services, openness and transparency and accountability (RSA 1997b). These did not preclude punishment but the emphasis on persuasive approaches makes the base of the regulatory pyramid seem far more 'Batho Pele'. EHPs also mentioned the 1997 *White Paper for the Transformation of the Health System in South Africa* whose environmental health chapter focused on five principles:

- Every South African has the right to a living and working environment which is not detrimental to his/her health and well-being.
- All persons should have access to knowledge on environmental health matters and the services available to them.
- Environmental health services should be accessible, acceptable, affordable and equitable. They must be implemented with the active participation of the communities.
- Environmental health services should contribute positively towards sustainable physical and socio-economic development.
- The establishment of effective environmental health surveillance is essential to determine whether or not the services are functional and effective and have a positive health impact.

(RSA 1997a)

Two points further articulate this environmental health vision. First, alongside its Constitutional first principle, it acknowledges that environmental health is a shared responsibility; here the environmental health sector itself is responsible for providing accessible services and supporting communities in managing their environmental health risks, but individuals remain ultimately responsible for maintaining a healthy environment (RSA 1997a). Second, the following statement under the heading 'environmental health legislation' provides a rare but significant clue for EHPs:

> A community development rather than a law enforced approach will be followed in creating environmental conditions conducive to good health. Environmental health legislation will comply with the requirements contained in the Interim Constitution's Bill of Rights and will be based on integrated, appropriate and uniformly applicable legislation.
>
> (RSA 1997a)

This was reinforced by the *National Environmental Health Policy for South Africa*, its first draft in 1996 calling on EHPs to adopt community development approaches and describing them as educators and facilitators, monitoring and responding to community needs and demands and enabling access to essential local authority services (Eales et al. 2002). By 2013 the key principles of its final draft encompassed the whole regulatory pyramid but a preference for persuasion remains:

> Although it's a requirement for compliance to national policies and standards on environment and health by importers, producers, manufacturers, retailers and communities, environmental health strategies should strike an appropriate balance between promotion and education and law enforcement … Education must be used as a vital tool of building capacity for all stakeholders/role players, to be able to attain voluntary compliance.
>
> (NDoH 2013:16)

What are environmental health services?

The *Constitution* lists 83 areas of legislative competence of national, provincial and local government in two schedules, with the Environmental Health Services of Urbington mainly categorized as a local government competence under the umbrella of 'municipal health services' but the term was beset by uncertainties. Whilst the *Constitution* mandates municipalities to deliver municipal health services, metropolitan municipalities like Urbington were not given legal responsibility for delivering them until the 1st February 1999 and what constitutes municipal health services (see Chapter 2) was not legally defined in the *National Health Act, 2003* until 2006.

Further, what's specifically included in these nine areas is not further defined but the Act excludes port health, malaria control and the control of hazardous substances which are designated provincial government functions (Agenbag and Balfour-Kaipa 2008). In response, the Professional Board for EHPs of the HPCSA was trying to provide some clarity, including a scope of the work of EHPs, but in 2006 this was little more than a long list of areas and sectors that ended with a catch-all sentence covering 'any other matter incidental to or of Environmental Health significance, which, if unattended would compromise the quality of public and Environmental Health' (HPCSA 2006a). In response to these uncertainties senior managers at Urbington confirmed that their provincial branch of the South African Local Government Association (SALGA) had pragmatically adopted a definition of municipal health services as environmental health and they had decided to deliver the nine functional areas whilst continuing to cover missing areas (e.g. accommodation establishments) and monitor premises run by other departments and provincial government (see Table 5.1).

Minimum legal standards and the inevitability of discretion

Great volumes of legislation[1] defined minimum standards and gives EHPs and others law enforcement powers but the law also conferred considerable discretion on them. Hawkins and Thomas argue that where regulatory goals are clearly defined and unambiguous they are likely to be enforced more stringently (1984) but environmental health law has been characterised by both broad and prescriptive standards since Victorian times (Crook 2007; Wilson 1881). In South Africa, the minimum legal standards for environmental health remained largely prescriptive and included requirements for permits or licenses to operate whose origins date back to the British colonial era (Rogerson 1986). For example, the *Regulations governing general hygiene requirements for food premises and the transport of food (R.918 of 30 July 1999)* (known to EHPs as the 'R918') prohibits certain activities, prescribes standards for food premises, equipment and handlers and requires a permit called a *Certificate of Acceptability* from the local authority in order to operate.

Similarly, Urbington's *Public Health By-laws* sets out the minimum space standards and facilities for various businesses (e.g. child care facilities) and require a permit for them to operate. Some permits (e.g. the Certificate of Acceptability for food premises) were free, but applications for others began with the full payment of fees alongside the collection of many details. Businesses were then subject to inspection by EHPs and others and only once all Departments were happy were permits granted. Under wider trading laws many businesses required a separate license to trade that involved the collection of details and the payment of a one-off application fee of around R150 ($21 in 2007) for informal premises or R500 ($70) for formal premises. Failure to comply with these requirements was an offence but permit processes were found to be ripe for exploitation by corrupt EHPs and others (see Chapter 7).

Prescriptive minimum standards typically co-existed alongside more general standards framed by terms like 'adequate', 'reasonable' or 'effective' that required the discretion and judgement of EHPs who sometimes spent considerable time trying to interpret the law. By linking the law to the offence in this way Hawkins (1984) argues that discretion gives the law its very substance and life. Whether discretion is problematic remains the subject of debate but Urbington's EHPs were subject to a range of controls, notably close supervision by managers as explored further in Chapter 5.

The uncertain criminality of environmental health offences

Mens rea, literally guilty mind, requires the prosecution to prove a defendant intended to commit a crime but across environmental health law it has largely been replaced by the principle of strict liability that mandates businesses to comply with legally prescribed standards. Strict liability emerged during the Victorian era to provide a deterrent whilst, in theory, making the law more enforceable (e.g. prosecuting inspectors no longer had to prove *mens rea*) though studies question whether this was achieved in practice (Carson 1979; Paulus 1974). There is some evidence that strict liability can make officials (Carson 1970a) and the judiciary (Ogus 2004) more hesitant to sanction offenders, but studies of EHPs and others suggest that *mens rea* remains an important factor in the decision-making of regulators (Carson 1970b; Hawkins 1984; Hutter 1988). Further, legal defences against strict liability continue to incorporate *mens rea*

(Carson 1970a), whilst Norrie (2001) adds that using the criminal law to regulate business activities is further limited by the challenge of establishing corporate criminal liability. For example, attempts to assimilate the criminal liability of the individual with those of complex organisations are fraught with problems like linking individual acts to those of the business itself, particularly where companies comprise many different structures and legal relationships (e.g. shareholders, contractors).

All these factors were playing out in Urbington and Chapter 6 explores EHPs' views on the criminality of offending but most South African law incorporates strict liability offences prescribing the minimum standards for businesses. Some laws incorporated *mens rea*, but this was mainly related to how standards are administered or legal defences, for example:

> A person is guilty of an offence if they obstruct or hinder a health officer, or refuse to provide them with information or knowingly provide them with false or misleading information.
>
> *National Health Act, 2003*

> An employer can be convicted for the offence of an employee or others unless he proves that he took all reasonable measures to prevent an act or omission of the nature in question.
>
> *Foodstuffs, Cosmetics and Disinfectants Act, 1972*

There was also some evidence to support Norrie's (2001) argument that the abilities of the criminal law to regulate business activities are limited by the challenge of establishing corporate criminal liability. EHPs explained that it sometimes took months, even years, to establish who owned premises or buildings under investigation because they were hidden by complex corporate structures, including shell companies. Some EHPs enjoyed the detective work involved here but others didn't, one commenting that it actually deterred him from taking on complex housing cases.

Wider questions have also been raised about whether strict liability offences are constitutional. For example, Lazarus et al. (1997) argue that strict liability offences might violate the fundamental right to a fair trial and thereby fail constitutional challenge unless provisions can be put in place to minimise the risk of convicting innocent people. There were no examples of this during fieldwork but two landmark housing cases revealed other tensions between environmental health law and Constitutional rights. For example, Huchzermeyer (2004) questions whether some contraventions of the law should instead be redefined as a lack of protection from breaches of Constitutional rights. In the case of *Government of the Republic of South Africa v. Grootboom, 2001* for example Irene Grootboom and her community moved from an informal settlement onto private land intended for low-cost housing in Cape Town but were then evicted and forced to live under plastic sheeting on a nearby sports field. Grootboom was successful in asking the Constitutional Court for the provision of adequate shelter/housing until permanent accommodation was available, but the Constitutional Court also ruled that State housing programmes were not reasonable because they failed to provide for those in desperate need.

In another example, Ray (2008) explores the case of *Occupiers of 51 Olivia Road v City of Johannesburg, CCT 24/07, Judgment of 15 February 2008* that involved a series of emergency applications to the Witwatersrand High Court by the City of Johannesburg Metropolitan Municipality to evict over 300 people from six bad buildings in inner-city Johannesburg. This was part of a wider regeneration strategy but the residents, assisted by a housing rights NGO and the local University, initially successfully opposed these applications on several legal grounds including the Constitutional right of the Municipality to provide access to adequate housing and the failure of its housing program to comply with its Constitutional and statutory duties. These judgements were partially overturned in a series of appeals, but the municipality's inner city eviction programme was also suspended and further concessions granted included improvements to the environmental health of existing bad buildings. In its final decision, the Court required a process of engagement to be followed whereby municipalities must consult with residents affected by policy decisions that may involve eviction to ensure their socioeconomic rights are protected. It was too soon for Ray (2008) to comment on its success, but this judgement sends out a clear, Constitutional signal away from offending and towards rights-based approaches that engage with civil society to work through these complex problems.

The variable powers of EHPs

The regulatory powers of local government EHPs are considerable and wide-ranging, for example, the *Criminal Procedure Act, 1977* authorises municipalities like Urbington to appoint EHPs as Peace Officers to enforce the law. The *National Health Act, 2003* authorises

EHPs to enter any premises, apart from private dwellings, at any reasonable time and it is an offence to obstruct or mislead EHPs in the conduct of their work. Such powers were first enacted more than 130 years ago but accounts from this period also describe widespread opposition to the expansion of surveillance of EHPs into formerly private spaces (Crook 2007). It was not uncommon for Urbington EHPs to be obstructed or misled in their work but individuals' rights to privacy are overridden here by concerns towards safeguarding public health. However, after Hutter (1988), the powers of Urbington EHPs can also be considered variable for two reasons. The legal right of access to a business does not automatically infer compliance with the law, for example, most food premises had facilities like wash-hand basins but food handlers were rarely seen washing their hands. Second, EHPs sometimes had little or no powers to regulate what was outside their responsibility, for example, Urbington's own premises and those monitored by provincial government (e.g. schools and prisons).

Sanctions for environmental health offences

The discretion of Urbington EHPs extends to the application of increasingly punitive sanctions. Chapter 6 explains that EHPs viewed criminal sanctions more as a tool towards legal compliance than to punish offenders but the maximum sanctions for single offences were published on Urbington's website and are summarised in Table 4.1.

Sanctions for multiple offences during the same visit were not uncommon, EHPs recalling spot fines totalling R5000 ($700) but for larger businesses, they knew such sums were unlikely to deter

Table 4.1 Maximum sanctions available to Urbington EHPs and/or the courts in 2007

Law	*Maximum sanctions in Rand ($ equivalent in 2007)*
Health Act 1977	R1500 ($210) imprisonment not exceeding 2 years; or both
Tobacco Products Control Act 1993	R1500 ($210)
Regulations governing general hygiene requirements for food premises and the transport of food (R 918, July 1999)	R500 ($70)
Provincial Noise Control Regulations	R1500 ($210)
Public Health By-laws	R1000 ($140)

offending and risked undermining their authority. The threat of imprisonment could be more effective, but no evidence of imprisonment for environmental health offences was found though the maximum period for imprisonment under the *National Health Act, 2003* was in the process of being increased from two to five years. In 2007 around 57% of South Africans were considered poor and earned less than R458 per capita per month (SSA 2014) but Urbington EHPs were also sometimes willing to sanction them.

Other sanctions included the prohibition of activities, the closure of premises and the withdrawal of licenses/permits but only more experienced EHPs were exploring these options, though work was underway to upskill EHPs here.

The uncertain role of the local government EHP

The precise role of local government EHPs was not clearly defined in law and across the national policy. The *Health Professions Act, 1974* requires EHPs to register each year with the Board for EHPs of the HPCSA and to confine themselves to practising in the field of environmental health. However, in 2007 this field was defined in the Board's 'Competency requirements and Scope of Profession of EHPs' in a long list of so-called matters relevant to EHPs, but no detail on what competent practice might look like (HPCSA 2006b). In 2007 the role of EHPs was better articulated in the *White Paper for the Transformation of the Health System in South Africa* and the *National Environmental Health Policies* that identified the need for approaches across the regulatory pyramid but expressed a clear preference towards persuasion. When asked to describe their own roles, most EHPs provided a technical response by reciting the nine areas that define 'municipal health services' and then adding 'anything else relevant to environmental health' before describing the typical premises and complaints they encounter. All acknowledged their law enforcement work but considered their primary role to be more about persuasion to protect people and improve their lives as explored in Chapter 6.

In the UK, Hutter (1988) suggests that the gradual development of local government environmental health 'more by chance than design' could explain why their legal mandate is so broad, complex and ill-defined. More recent historical work provides some support for this, notably Crook's (2007) exploration of the myriad factors that drove the expansion of the law and the role of British EHPs from the late Victorian period. South Africa's history suggests this expansion was more deliberate given the role of EHPs in maintaining and expanding colonialism and apartheid but Urbington EHPs raised important concerns about the law and its suitability for regulating local environmental health that is explored further in Chapter 6.

The statutory position of EHPs

Legal responsibilities for environmental health policy are fragmented across national, provincial and local governments. In theory, this shouldn't be a problem given that the *Constitution* designates all three spheres of government as equal and interdependent and obliges them to cooperate with each other but studies suggest the national policy influence of EHPs is limited. The Environmental Health Directorate of the National Department of Health has responsibility for policy making, coordination and support for provincial and local government EHPs (NDoH 2006) but this is challenging when responsibilities are cut across so many government departments as summarised in Table 4.2.

Senior EHPs and others (Agenbag 2006; Haynes 2005) voiced concerns about the lack of national leadership, not helped by long periods without a National Director of Environmental Health in post. In 2007 the Board for EHPs of the HPCSA and the South African Institute of Environmental Health (SAIEH) were trying to influence national policymakers but by 2013 it was clear that they and the National Director of Environmental Health, by then an EHP and former secretary of the SAIEH, were becoming more established advocates for EHPs. In Urbington itself managers were working to reduce fragmentation and duplication in the delivery of their environmental health services but with reorganisation ongoing, it was unclear how this would progress.

For EHPs themselves the *Health Professions Act, 1974* provided an important statutory basis for their work alongside the duty of EHPs to be registered with the HPCSA to practice. However, EHPs also thought the HPCSA could do more to protect and promote their work, including from a perceived threat of Environmental Health Assistants and new Environmental Management Inspectors (focused on environmental pollution control) taking their jobs. There was some legal basis for these fears because most laws do not specify who is responsible for their enforcement and refer instead to general terms like 'inspector' or 'authorised person'. In 2007 responsibility for the delivery of municipal health services remained, by definition, within local government but EHPs themselves were increasingly working for other local government departments, in provincial and national

Table 4.2 National and Provincial Government Departments with EH responsibilities

National Departments	*Scope of EH responsibilities*
Environmental Affairs and Tourism	Chemicals and waste management policy Environmental advisory services & programmes Climate change and air quality policy
Housing	Housing and sustainable human settlements
Labour	Regulating occupational health and safety
Minerals and Energy	Regulating mines health and safety. Environmental management of mines and other energy sources.
Trade and Industry	EH standards policy – e.g. South African Bureau of Standards
Transport	Regulating transport safety
Water Affairs and Forestry	Water & sanitation policy Water resources management
Provincial Departments	*Scope of EH responsibilities*
Agriculture, Conservation and Environment	Regulating waste management, air quality and environmental management
Community Safety	Projects related to safety in schools and communities including alcohol, drugs
Education	Provision of school meals (inc. nutrition & food safety)
Health – Environmental Health Services	Regulating provincial premises, port health, control of communicable diseases, hazardous substances, malaria control, environmental health sampling (e.g. food, water) Coordination functions – overseeing activities of Provincial local government, including the Urbington.
Health – wider public health	Delivery of primary health care services (inc. HIV/AIDS and TB) across hospitals, clinics and in the community.
Human Settlements	Housing and sustainable human settlements in coordination with local government
Infrastructure Development	Development and maintenance of provincial environmental health-related buildings and infrastructure
Roads and Transport	Sustainable transport policy
Social Development	EH-related services for vulnerable groups (e.g. disabled, elderly, those affected by HIV/AIDS)

government and the private sector. Given the fragmented nature of environmental health regulation and the continued influence of neo-liberal policies on local government, it remains quite possible that in the future EHPs could find themselves working for more centralized, or privatized, services.

Note

1 EHPs maintained a working knowledge of more than 20 Acts of Parliament, numerous national and provincial Regulations and standards and Urbington's own By-laws.

References

Agenbag, M. (2006) An analysis of progress made with the devolution of municipal health services, in South Africa Unpublished conference paper presented at SAIEH conference, eThekwini, South Africa, February 2006.

Agenbag, M. and T. Balfour-Kaipa (2008) Chapter 10: Developments in environmental health, in Barron P. and Roma-Reardon J. (Eds) *South African Health Review 2008*. Health Systems Trust, Durban, South Africa.

ANC African National Congress (1994) *The Reconstruction and Development Programme: A Policy Framework*. African National Congress, Johannesburg, South Africa.

Carson, W. G. (1970a) Some sociological aspects of strict liability and the enforcement of factory legislation, *Modern Law Review*, July 1970 396–412.

Carson, W. G. (1979) The conventionalization of early factory crime. *International Journal of the Sociology of Law*, Vol. 7 37–60.

Carson, W. (1970b) White-collar crime and the enforcement of factory legislation *British Journal of Criminology*, Vol. 10 383–398.

Coovadia, H., R. Jewkes, P. Barron, D. Sanders and D. McIntyre (2009) The health and health system of South Africa: historical roots of current public health challenges. *The Lancet*, Vol. 374 (9662) 817–834.

Crook, T. (2007) Sanitary inspection and the public sphere in late Victorian and Edwardian Britain: a case study in liberal governance. *Social History*, Vol. 32 (4) 369–393.

Eales, K., S. Dau and N. Phakati (2002) Chapter 6: Environmental health. *South African Health Review 2002*. Health Systems Trust, Durban, South Africa.

Habib, A. and V. Padayachee (2000) Economic policy and power relations in South Africa's transition to democracy. *World Development*, Vol. 28 (2) 245–263.

Hawkins, K. (1984) *Environment and Enforcement: Regulation and the Social Definition of Pollution*. Clarendon Press, Oxford, UK.

Hawkins, K. (2002) *Law as Last Resort: Prosecution Decision-Making in a Regulatory Agency*. Oxford University Press, Oxford, UK.

Hawkins, K. and J. Thomas (1984) The enforcement process in regulatory bureaucracies in K. Hawkins and J. Thomas (Eds.) *Enforcing Regulation.* Kluwer-Nijhoff, Boston, USA.

Haynes, R. (2005) *Monitoring the Impact of Municipal Health Services Policy Implementation in South Africa.* Health Systems Trust, Durban, South Africa.

Hood, C. (1991) A public management for all seasons? *Public Administration,* Vol. 69 3–19.

HPCSA Health Professions Council of South Africa (2006a) *HPCSA EHP competency requirements and scope of profession.* HPCSA, Pretoria, South Africa.

HPCSA Health Professions Council of South Africa (2006b) *Introducing the Professional Board for EHPs.* HPCSA, Pretoria, South Africa.

Huchzermeyer, M. (2004) From 'contravention of laws' to 'lack of rights': redefining the problem of informal settlements in South Africa. *Habitat International,* Vol. 28 333–347.

Hutter, B. (2005) *The Attractions of Risk based Regulation: Accounting for the Emergence of Risk Ideas in Regulation.* ESRC Centre for Analysis of Risk and Regulation Discussion Paper No: 33 London School of Economics, London, UK.

Hutter, B. (1988) *The Reasonable Arm of the Law? The Law Enforcement Procedures of Environmental Health Officers.* Clarendon Press, Oxford, UK.

James, P., S. Tombs and D. Whyte (2013) An independent review of British Health and safety regulation? From common sense to non-sense. *Policy Studies,* Vol. 34 (1) 36–52.

Lazarus, P., I. Currie and R. Short (1997) Chapter 1 The legislative framework: environmental law, investment and industrial practice, in Bethlehem, L. and M. Goldblatt (Eds.) (1997) *The Bottom Line: Industry and the Environment in South Africa.* International Development Research Centre (Canada) and University of Cape Town Press, South Africa.

NDoH National Department of Health (2006) *Environmental Health* Accessed on 27 October 2006 via: http://www.doh.gov.za/department/dir_environthealth.html.

NDoH National Department of Health (2013) *National Environmental Health Policy.* Government Gazette No. 37112 4 December 2013.

Norrie, A. (2001) *Crime, Reason and History: A Critical Introduction to Criminal Law.* Cambridge University Press, UK.

Ogus, A. (2004) *Regulation: Legal Form and Economic Theory.* Hart Publishing, Oxford, UK

Osborne, D. and T. Gaebler (1992) *Reinventing government: how the entrepreneurial spirit is transforming the public sector* Plume Books, Penguin Books, New York, USA.

Paulus, I. (1974) *The Search for Pure Food; A Sociology of Legislation in Britain.* Martin Robertson & Company Ltd., London, UK.

Ray, B. (2008) Occupiers of 51 Olivia Road v City of Johannesburg: Enforcing the right to adequate housing through "engagement". *Human Rights Law Review,* Vol. 8 (4) 703–713.

Rogerson, C. M. (1986) Feeding the common people of Johannesburg, 1930–1962. *Journal of Historical Geography*, Vol. 12 (1) 56–73.

RSA Republic of South Africa (1996) *Growth, Employment and Redistribution: A Macroeconomic Strategy* Government Printer, Pretoria, South Africa.

RSA Republic of South Africa (1997a) *White Paper for the Transformation of the Health System in South Africa*. Department of Health, Pretoria, South Africa.

RSA Republic of South Africa (1997b) *White Paper on Transforming Public Service Delivery (Batho Pele White Paper)*. Department of Public Service and Administration, Pretoria, South Africa.

RSA Republic of South Africa (1998) *White Paper on Local Government* Department of Constitutional Development, Pretoria, South Africa.

SSA Statistics South Africa (2014) *Poverty Trends in South Africa: An Examination of Absolute Poverty between 2006 and 2011*. Statistics South Africa, Pretoria, South Africa.

Tombs, S. (2016) *Social Protection after the Crisis: Regulation without Enforcement*. Policy Press, Bristol, UK.

Wilson, F. (1881) *A Practical Guide for Inspectors of Nuisances*. Knight & Co., London, UK.

5 Urbington Metropolitan Municipality strategies and controls

Urbington's strategic framework

Urbington's own strategies were bringing the legal framework for local government and environmental health from Chapter 4 to life in many ways as encompassed by its 2030 vision:

> Urbington will be a city of opportunity, where the benefits of balanced economic growth will be shared in a way that enables all residents to gain access to the ladder of prosperity, and where the poor, vulnerable and excluded will be supported out of poverty to realise upward social mobility. The result will be a more equitable and spatially integrated city … [where] everyone will be able to enjoy decent accommodation, excellent services, the highest standards of health and safety, and quality community life in sustainable neighbourhoods and vibrant urban spaces.

Environmental health cut across Urbington's strategies but also gave mixed messages to EHPs on whether to persuade or punish. For example, one referred repeatedly to 'combating pollution using rigorous law enforcement', another emphasized persuasion via 'long-term strategic interventions' and working together with businesses and communities, with law enforcement only 'where necessary'. Another sought to establish a citywide approach to Bylaw enforcement in order to prevent 'unnecessary and prohibitive fining' when more persuasive approaches could be more effective.

Monitoring work for others

Figure 3.1 introduced the monitoring role of EHPs for other Urbington organisations and provincial government with typical roles

DOI: 10.1201/9781003009931-5

Table 5.1 Examples of environmental health monitoring by EHPs for others

Organisation	*Examples of monitoring activities*
Urbington Agencies	
Urbington Roads	Vacant land conditions and roads (e.g. blocked drains)
Urbington Parks	Vacant land conditions and parks
Urbington Departments	
Emergency Management Services	Joint approval of permits for pre-schools, advice on fire safety and water safety issues
Health	Immunisation programmes in schools and crèches
Housing	Water and sanitation in informal settlements Environmental health conditions in residential buildings
Transportation	Diesel vehicle emissions
Planning and Development	Location and use of buildings and settlements
Environment	River and drinking water quality
Urbington Utilities	
Urbington Waste	Waste management
Urbington Water	Water and sanitation (e.g. leakages, overflowing drains)
Provincial government	
Environmental Health Services	Environmental health of provincial government schools, prisons, health clinics

summarised in Table 5.1. This work was all advisory and EHPs described generally good relations with the organisations they monitored for and how this work developed their knowledge for the benefit of others as summarised by one senior manager:

> … if you pick up water running down the street, it might not even be a health nuisance, but there's water running so you report it to Urbington Water. There's nothing preventing you from contacting Urbington Roads if you find that there's a pavement that's got a pot hole in it or whatever the case might be, and you as an EHP might think that this is a safety issue for pedestrians.

> It's got nothing to do with you directly, it's not your core business, but YOU are the eyes and ears of Council out there.
>
> (Assistant Director – Environmental Health)

Other EHPs considered working with so many organisations to be very much part of their role and it provided many opportunities, including identifying premises missing from databases. For example, citywide immunisation programmes included the identification (and subsequent regulation) of many previously unknown crèches. Such joint monitoring working could also benefit these businesses, with EHPs forwarding copies of crèche permits they processed to Urbington Community Development to support social grant applications by the crèches to improve facilities.

However, EHPs had concerns about monitoring work, particularly when they thought other organisations were responsible or a case went beyond the scope of their work:

> … your correctional services, your police stations, your clinics, your schools … are supposed to have either their own health inspector, or the provincial government inspectors must go there. But then we do them. We must tell Urbington Waste that there's dumping somewhere, and that's the only time they go and pick it up. We must tell Urbington Parks that there's overgrown grass. Then that's when they go and cut it … when I see a leaking water pipe I call Urbington Water …
>
> (Township EHP 2)

> … it's not part of our job to be involved in immunisation. Our job is to make sure that there are measures to control communicable disease, or investigate communicable disease, but not actually involve in the immunisation. But people think that communicable disease is one of our field so we have to be involved in immunisation …
>
> (Inner city EHP 1)

Some EHPs felt further demoralised when monitoring activities weren't included in their performance targets and reduced their time for core work, a situation one EHP likened to being robbed. Some also questioned why other organisations were not doing their own monitoring and feared such arrangements set a precedent that could make matters worse:

> I think the problem is WE as the Health Department, people come, people expect US to do the work for all of the other

> Departments, I think that's the main thing. We must report on about everything that goes wrong, if the storm water drains are not working, if the lights are not working, if there's problems with overcrowding, town planning, I mean if you think it's pushed up to us, we must send the things through to them.
>
> (Inner city EHP 4)

> Ok, you're already there, tell us where they are dumping, you're already there, you can tell us if it's overgrown, you're already there, you can tell us if there's an invasion of land, you can tell us of the leaking pipe, you can tell us where there's building rubble …
> (Township EHP 2, imitating another Department speaking to her)

> We are the only street-level people in the Council … [Urbington Waste] don't have law enforcement people, Environmental Management don't have law enforcement … We are the only ones that can go out and do real inspections in the streets … the other Departments don't have and then they piggy back on us to do their things. And they will take all the credit for it!
>
> (Inner city EH Manager)

EHPs were also concerned that their monitoring reports were sometimes acted upon slowly, if at all. Knowing who to speak to was critical but EHPs provided many examples of organisations taking a long time to respond to reports of serious and persistent problems. When there was no response EHPs felt frustrated and powerless, particularly when they revisited and were then blamed by the public and media for continuing problems. The failure of some Urbington organisations to maintain their own environmental health standards could further complicate matters. For example, residents in one region were complaining about water entering their newly built homes. EHPs initially discovered inadequate drainage but wider investigations found they were built on a flood plain, something EHPs thought easily preventable if Urbington Planning and Housing were doing their jobs. In another region the failure of Urbington to maintain its own land was making neighbours respond accordingly:

> … you [Urbington] say I must clean my own house, you clean yours first.
>
> (Suburban Operations Manager)

But EHPs also observed that some Urbington organisations were gradually taking responsibility for environmental health matters.

For example, EHPs were increasingly referring pollution cases to Urbington's new Department of the Environment, waste management cases EHPs had traditionally handled were forwarded to Urbington Waste. There was also a growing workforce of Environmental Management Inspectors, including some EHPs, but instead of welcoming these developments, some EHPs considered them a threat to their own roles and professional status.

Joint working on permits/licenses with other Urbington organisations could also be problematic. Whilst the law prohibited businesses from trading without a permit/license in Urbington people continued to trade and open new businesses without them:

> If we find somebody trading we cannot close those people down. Instead we have to say 'please come and apply for a license'. They will take their time, two or three months down the line you have to send out a notice, even summons them, but then speak to them first. Finally they come to apply for a license, you have to still wait for these other Departments to give their ok before you give the license. These are the ones that takes their time ... there's nothing you can do. You cannot close the school down because you are still waiting for the Fire Department ... You have to chase after them just to get a report back ... a license application may take up to four or five years ... You end up chasing and fighting with your own colleagues here, it's just ridiculous.
>
> (Suburban Operations Manager)

Another example involved an ongoing bat infestation in a block of flats in which the Departments of Health and Housing were both refusing to carry out the specialist works following quotes of R100 000 (~$14000 in 2007). However residents continued to complain and bat faeces covered the external walls, corridors and doors. Working with Urbington Housing it could take EHPs 3 years or more to obtain a Demolition Order for a bad building and in the meantime, residents had moved back in.

Organisations were also prepared to challenge EHPs on their monitoring. For example, Suburban EHP 3 recalled being contacted by an Urbington Housing official after he wrote a report about an informal settlement. The official questioned who he was to say such things, and he replied by stating that he was telling the truth and the official put the phone down on him. He reported the incident to his managers whose actions led to a meeting the following week at the informal settlement with community leaders and Urbington officials who validated EHP 3's report. Attempts to recover costs by EHPs were often challenged, particularly for pest control work, when the following replies were typical:

> No, we don't have money for that, we are the Council, you are the Council ...
>
> (Suburban Operations Manager imitating another Urbington Department)

Monitoring work could also harm relations between EHPs and the public:

> ... it always looks like we are spying on communities, and then giving other Departments information for forceful action like removals, taking of their possessions and evictions.
>
> (Township EHP 2)

Similarly, EHPs knew their monitoring reports were used to condemn buildings as unsafe and the consequences made them very uncomfortable:

> ... here in the City centre, there's bad buildings, there's evictions and they [other Urbington Departments] want EHPs to be there, to be present when actual evictions are taking place, of which we have done our part, we have inspected and said the building it's unsafe, or unhealthy for people to stay there, we've done the report. They have the powers to take action, but they want us to be there on the day of the eviction, of which it is not our role to be evicting the people, or to make sure that the people are eventually evicted.
>
> (Inner city EHP 1)

EHPs also commented on how the reorganisation of Urbington was contributing to these problems, particularly when what they called 'shifts in power' made the avoidance of responsibility easier. But EHPs were also hopeful things would improve. For example, new licensing policies would soon ensure planning controls (e.g. zoning and acoustics) were adequate before nightclubs were allowed to open. The Department of Health was also exploring legal options to force the Department of Housing to improve serious and persistent sanitation problems in one informal settlement. EHPs were also united in their unhappiness monitoring the provincial government and had their own EHPs. In the meantime, they were exploring options for law enforcement against provincial government premises but senior managers were less keen and stressed that, when dealing with other state bodies, all possible options should be considered before going to court.

When asked why these monitoring problems persisted, EHPs typically identified shortages of staff and other resources and accused some organisations of silo working. This was verified by senior managers and politicians, the former also conceding that some organisations routinely avoided their responsibilities for environmental health and therefore EHPs were playing a vital role in trying to hold them to account. Therefore whilst sometimes beneficial, the costs of monitoring work on Urbington's EHPs were also very real.

Relations between EHPs and managers

All EHPs spoke of general conformity in regulatory approaches between their colleagues that met with the expectations of their managers and began during induction. This usually started with an orientation meeting with new colleagues followed by a tour of their areas with their Operations Manager to identify boundaries and potential priorities or concerns. Inner city EHP 2 likened his first tour of Urbington to 'another page' opening up for him and being shocked by the 'real picture' he saw. Indeed, one Suburban manager stressed the vital support needs for his new EHPs:

> … [they] need to be built up, need to be trained, need to be literally taken by the hand and shown 'this is the boundary of your Region, your District' … 'this is what a food shop looks like, this is what you should be looking for, where you find a problem, this is how you talk to a person, this is what you're entitled to ask for, if they don't do it, this is the process you follow …
>
> (Suburban EH Manager)

Managers worked closely with new EHPs until they became familiar with their areas, the job itself, the law and the paperwork, a process that could take 3 months or more. Colleagues also provided vital support during this period, particularly on cases and paperwork, though the lack of experienced EHPs to mentor new recruits was an ongoing concern. EHPs all described good relations with their colleagues and would often ask each other for advice and share ideas and information, a process helped by the open plan layout of most offices. Sometimes EHPs worked together in the field, particularly when they had safety concerns (see Chapter 6) or needed help with translation or less pleasant tasks (e.g. inspecting mortuaries).

It was not uncommon for EHPs to eat lunch together and many socialised together outside work, some even lived together. They also had good working relations with their Operations Managers and considered them to be generally supportive, good sources of information and advice and treated them as professionals.

There were also tensions between EHPs and their managers, most due to what one termed 'differences of opinion about priorities'. EHPs often complained about pressures to meet performance targets but they sometimes accused managers of being out of touch with their areas, particularly in areas like personal safety. EHPs also disliked managers imposing work on them, particularly at short notice:

> I have a campaign and I was just told about it this week: 'next week it's gonna be a campaign, and it's gonna go like this.' We EHPs are not involved in the planning. Things are just done, they come from on top, because in EH we know exactly what is going on, we know if we come with this approach it's not gonna work because we know those people, we work with those people … So maybe it's one of the reasons why campaigns or projects don't succeed, because the people on top come and impose things.
>
> (Inner city EHP 1)

EHPs knew their managers were in charge and were under pressure from senior managers but they also did not think they should be blamed when projects like this failed. Instead, EHPs wanted to be more involved in developing projects from the earliest planning stages and some senior managers agreed that front-line EHPs should be more involved in developing Urbington's future environmental health policies.

Two vignettes based on actual cases were developed to further explore departmental conformity and expectations and are summarised in Table 5.2. In both scenarios, most EHPs thought their colleagues would do the same and that this was as per their managers' expectations. Their approaches were also similar and favoured persuasion in the first instance, escalating up the regulatory pyramid if unsuccessful. However, some EHPs were unsure what their colleagues would do or feared they might act differently. The experience was key here, where an inexperienced EHP might issue a permit in the informal tuck shop vignette or, at the other end of the pyramid, issue a spot fine or close it down. For the crèche

Table 5.2 Vignettes exploring practice, organisational conformity and expectations

1 Imagine that while driving through a remote part of your area you notice a new tuck shop serving an informal settlement. There are no other tuck shops in the area. On your first inspection, you find that the tuck shop is built of temporary materials and sells only packaged bread and a few tins of food.

a How would you proceed in this situation?
b What do you think your EHP colleagues would do?
c What would your Managers expect you to do?
d Would you act differently if it was the last day of the month and you had only served two fines instead of your monthly target of three?

2 A crèche applying for a permit in your area has been given provisional approval by the Emergency Services, Building Department and Licensing inspectors. The person providing the service is friendly and cooperative and is providing facilities and care of a high standard, however, your calculations of indoor care area space find that there is only space for 30 children, but the crèche has 35 each day.

a How would you proceed in this situation?
b What do you think your EHP colleagues would do?
c What would your Managers expect you to do?
d Would you act differently if the service provider was aggressive and uncooperative?

vignette, an inexperienced EHP might allow the crèche to continue operating whilst overcrowded and one EHP admitted to making this exact mistake in the past.

For both vignettes, almost all the EHPs thought their managers would expect them to follow their chosen courses of action. Only one, Inner city EHP 4, thought his manager would expect a more punitive approach, what he characterized as 'no negotiations, zero tolerance', but he also respected his manager's right to see things differently. Similarly, most other EHPs thought cases like these were not straightforward and required considerable discretion because, as one put it:

> ... sometimes you can't stick to the law 100%, we need to be practical.
>
> (Suburban EHP 1)

Further, most EHPs added that in such circumstances they would also brief their manager and seek their advice and endorsement before taking further action.

Performance management controls

The working day itself provided a basic control on the activities of Urbington EHPs, particularly the expectation that they would work in the field from 10:00 AM until around 15:00 PM unless they had good reason and management permission. Their workloads were also influenced by the nature of the job, particularly the number of complaints received, and the influence of their colleagues and managers as explored earlier. But Urbington's performance management systems effectively reinforced the regulatory pyramid and were introduced in 2000. They reflected the city's strategic priorities and were based on Kaplan and Norton's (1996) balanced scorecard, where 'balance' refers to the use of non-financial indicators to measure performance like those related to employees, activities and customer satisfaction. For example, the monitoring tools for EHPs comprised 65 key performance indicators with monthly and yearly targets for Urbington and its Regions across four main areas:

- Type and number of premises inspected
- Type and numbers of law enforcement & project related actions
- Number of samples/tests conducted
- Number of requests for service attended to, resolved or referred

Performance data were also collected relating to permits/licenses (e.g. the number of permits issued to formal and informal food premises) and to monitor persistent offending (e.g. the number of spot fines issued to the owners/occupiers of buildings that have failed to comply with notices).

The annual targets for Urbington for 2 years and one typical EHP conformed to the shape of the regulatory pyramid as summarised in Table 5.3. Yearly targets were broken down by the regions into monthly/weekly targets and allocated to EHPs and others through yearly performance agreements based on their responsibilities and areas. Each EHP then maintained a daily scorecard of their activities that was compiled into a weekly scorecard for their Operations Manager. The scorecards for EHPs differed but expectations of activity were similar and could increase by up to 10% year on year.

Table 5.3 Annual targets for Urbington environmental health and a typical district EHP

Activity target	*EH targets for 2005–6 and 2006–7*	*Typical EHP targets – per year*
Number of spot fines	2495	36
Number of statutory notices	29 781	264
Number of premises visited and evaluated/ re-evaluated	No target data, but 122262 evaluations over 2005–6	1800 (~8 per day)

All EHPs thought their work should be monitored, but most were ambivalent about targets:

> on the one hand they are something you cannot go away from, but on the other they can usually be met without problems.
>
> (Inner city EHP 4)

Targets were important, particularly for inexperienced EHPs, and their constant review by managers helped to verify their work and protect them from criticism. However, most EHPs thought their work was monitored too much. EHPs believed the time to complete scorecards could be better spent in their areas and EHPs undoubtedly felt pressured at times, particularly when dealing with lots of complaints or complex, longer-term cases. If they failed to meet their monthly targets EHPs had to provide reasons in writing to their Operations Manager. Missed targets could be carried over to the following month but the continued failure to meet targets could prompt disciplinary action but this didn't seem to happen very often.

Some activities (e.g. blitz work targeting offending hot spots) also meant that EHPs could sometimes meet a monthly target (e.g. 3 spot fines) in one evening. EHPs also made agreements with their Managers in response to local issues, one suburban EHP chose to inspect all the daycare centres and crèches in his areas monthly, not quarterly, to maintain standards there. Though EHPs invariably exceeded their targets there was little incentive to do so:

> ... even if you exceeded FAR beyond what they actually put they don't even look, there's not even say 'ok, fine, well done, so and so has done EXTRA work or double work per month.' No, there's nothing, you did what you're supposed to.
>
> (Inner city EHP 2)

EHPs were also concerned that targets could distort their decision-making in unfair ways:

> Maybe the person deserved the fine … maybe they were breaking the law and get a fine, but was the fine the right thing to do on that occasion? It's been determined by the target, not by what's required.
>
> (Suburban EH Manager)

> … you can't teach a person today, and then tomorrow you fine him. And then tomorrow you want to teach him again something … by then … there's that wall, people no longer listen to you open mindedly. They are like 'Ok, my sister …,' maybe they're going to fine me …
>
> (Township EHP 2)

> … maybe you told that person, maybe once, that they must provide a hand wash basin. You get there the next time, maybe they provided something to that effect, but not fully to your satisfaction. Under normal circumstances you will tell that person 'No, ok, you have put a hand wash basin but it's only cold water and I said I wanted hot and cold water' … and under normal circumstances maybe you would write another notice or you would be able to say 'no, put hot and cold water.' But then because a fine is wanted, you fine that person for not providing hot water … it causes tension because that person is TRYING to comply with your requirements but you are also trying to reach your targets. So sometimes it puts undue pressure on us and then we put undue pressure on the clients.
>
> (Township EHP 2)

> We've always done prosecutions, we've always done [spot fines], I'm the kind of person that says you can't get away from it, but it needs to be the end of the process, towards the end of the process. Targets unfortunately makes it sometimes the beginning.
>
> (Suburban EH Manager)

To examine this further the informal settlement tuck shop vignette in Table 5.2 asked EHPs if they would act differently if it was the last day of the month and they had only served two spot fines instead of their target of three. Six stated they would not act differently, particularly on their first visit, one commenting:

> I'm not writing tickets because 3 per month is required ... There must be justification, not just a chase for the scorecard.
>
> (Inner city EHP 4)

For the other four EHPs the target was important but the spot fine was a last resort and all agreed that they could be counterproductive for informal sector cases like this, unless there was a serious public health problem. Others feared such action could be seen as vindictiveness and damage their relations with businesses and the public, particularly if EHPs were thought to be raising revenues in this way. There were no revenue targets in the scorecards but this was clearly being considered because some included an 'income generated' box.

EHPs were also concerned that targets did not reflect the value of their work. Township EHP 2 commented that she was 'not implementing environmental health as a whole' because many core issues in her areas were not targets. Inner city EHP 1 commented that instead of being measured by the changes you've achieved a culture of 'pleasing your boss' by meeting your targets now existed as echoed by one Manager:

> If you keep on telling [EHPs] to do just this and this and this, unfortunately the score card creates gaps, because people end up just focusing on the score card. People forget about all the other 80% of environmental health functions ... So the score card has created a monster which is causing these EHOs to lose the skills they've developed on other things because in the end, in 5 years time, they will only focus on those things, they will lose the sight of the other things, that's what I'm afraid of.
>
> (Inner city EH Manager)

> [Targets] tell you this is what you must do, this is your [scorecard] for the year, do it. So you act like a robot .... [EHPs] lose their - 'I want to do something more, I want to do this and I want to look into this' because I don't have time to look into this because I must write so many stat. notices, so many fines' – lose their creativity, I'm worried about it.
>
> (Inner city EH Manager)

Another manager feared that some EHPs were becoming 'negative' by focusing solely on targets.

Other issues were creating further complications. EHPs on permanent contracts who met their targets typically got up to 3 days extra holiday per year or permission to attend a conference. But EHPs on temporary contracts could be eligible for financial rewards of up to 10% of their salary, whilst senior managers received financial bonuses. One manager, on condition of anonymity, commented that this was creating a corrosive target culture divorced from environmental health realities:

> [As an EHP] you work for somebody's bonus, you don't work for improving the quality of the environment. And then you'll get the Mayor in an email telling you: 'All managers please resign because their areas are not what they're supposed to be!' But in July, all the executives got their bonuses, so how can you tell us the areas are clean, they got their bonuses.
>
> (Anonymous EH manager)

EHPs and managers, therefore, agreed that the balanced scorecard system needed reviewing, including the development of targets based on outcomes, not activities. For example, one manager suggested targets based on compliance:

> I would like to have a target: 'how many of your crèches have permits?' SIMPLE! Not how many visits are you doing per year, but how many of them comply with the By-Laws … To say at the end of the year we have 50% comply, by now we have 70% complying, then you can say we've moved 20% or something like that …
>
> (Inner city EH Manager)

Similarly, for vehicle testing, EHPs suggested targets based not on the number of vehicles tested but on the numbers that pass/fail the emissions tests which could also enable better targeting of polluting vehicles. A few compliance-related targets (e.g. numbers of permits issued to formal/informal food premises) already existed and EHPs hoped more changes like this could help them evidence their work but past reviews with senior managers making similar recommendations had been unsuccessful.

Relations between EHPs and politicians

Urbington's most senior politicians regularly gave their support for and commitment to environmental health in many different ways. In policy terms environmental health was embedded in the Mayor's priorities alongside wider commitments to provide access to clean water and sanitation for all communities by 2010, to expand air quality monitoring and rehabilitate illegal dumping hotspots. Further commitments included the employment of more EHPs and continued public awareness campaigns on food handling and chemical storage. The Councillor leading Urbington's Department of Health admitted she did not know much about environmental health before being appointed but was now the 'leader of rat chasers' and she promised to increase their budget by R5 Million (~$700 000) the following financial year. Further, EHPs agreed that the Councillor had been very supportive of their work to date and had not sought to interfere.

Most front-line EHPs were engaged with the Ward Councillors for their areas, though one Township EHP admitted to having never spoken to a politician in the last 5 years though she knew they shared a regional office. Ward Councillors were a constant source of information and complaints and an important means of detecting local issues and potential offending. Working with them also helped EHPs access local communities via meetings or projects, though sometimes relationships could be very one-sided:

> … we hear from [Ward Councillors] once only if they need the help from us, whenever we need the help from them they are not there to phone.
>
> (Inner city EHP 2)

EHPs also helped inexperienced Ward Councillors navigate how the council worked, sharing their knowledge of its many services whilst some managers commented how this reflected well on EHPs (e.g. demonstrating the breadth of their knowledge), they also thought some asked for too much and instead needed more training and support here.

EHPs rarely felt pressured by politicians in their day jobs but all acknowledged that they regularly held them to account for their work and sometimes sought to influence it. Occasionally politicians told EHPs what to do, as was observed during one noise complaint investigation when an EHP was told by a Ward Councillor not to contact a complainant. Similarly, one manager reflected

> If we haven't done our work, wrap me over the knuckles [Ward Councillor] and say 'you haven't done it, I'm not happy, WHEN can I expect you to do it?' But don't tell me how to do it AND, don't hold ME responsible for the whole of council's work
>
> (Suburban EH Manager)

This was illustrated during tours of Urbington by senior politicians during fieldwork. One manager likened such tours to a 'wake up call' for politicians, particularly one by the Mayor that retraced a previous tour 5 years before and found no improvements in some areas. On this occasion, the Mayor concluded that his managers and front-line officials were 'dead', though EHPs also reflected that politicians sometimes didn't appreciate the scale and complexity of what they faced in Urbington. Managers also wanted politicians to think beyond their 5-year terms, engage more with the people and enforce the law, one commenting anonymously:

> People see illegal [electricity] connections, but nothing is done about them. People are invading factories, but what are officials doing? The community think it's a lawless society!
>
> (Anonymous EH manager)

EHPs could also find themselves caught up in local party politics, particularly when Ward Councillors made undeliverable environmental health promises to people during elections. Another tactic during election periods involved opposition Ward Councillors trying to show EHPs are doing nothing in order to send a message to the people that if you 'put us in power next time, we will do x, y and z'.

Resources for environmental health services

The realities of resource shortages were enshrined across the law, the *Constitution* for example requiring municipalities to strive to deliver their objectives '... within their financial and administrative capacity' whilst Urbington's own *Public Health By-laws* incorporated a similar clause:

> Every person has a constitutional right to an environment that is not harmful to his or her health or well-being ... and the Council has a constitutional duty to strive, within its financial and administrative capacity, to promote a safe and healthy environment.

In 2003 Haynes (2004) found that over 2002–3 the average spend on core local government environmental health services in South Africa was nearly R9 (~$1.3 in 2007) per capita though Urbington was spending nearly R14, just exceeding the National Department of Health's suggested per capita spend of R13. 3 years later Urbington's budget had increased to R14 per capita which covered around 20% of the total Environmental Health budget, the remaining 80% coming from Urbington itself and mostly covering salaries. This central budget also provided regions with the same basic resources, though regional managers submitted additional yearly budget requests to cover core functions and other activities like project work. Urbington was therefore relatively well funded compared to other municipalities but provided further evidence for Agenbag's (2006) argument that when compared to the national budget for primary health care services of R150–260 (~$21–36 in 2007) per capita, preventative services remained the poorer relative to curative ones.

The national government could intervene if a municipality was in financial trouble, an important safeguard because South African local government has historically been largely self-sufficient and raises around 90% of its own revenues (Beall et al. 2002). The work of Urbington EHPs raised some revenues (e.g. charging for some permits/licenses) but these would barely cover the costs of the time taken to administer them, let alone run a service. There was no evidence of 'revenue raising' pressures being put on EHPs though the public sometimes suspected this was happening, particularly with sanctions like the spot fine.

Ratios of population per EHP provide another crude way to explore resources. Haynes (2004) found a national average of nearly 24000 people/EHP in 2003, only the Western Cape meeting the National Department of Health (NDoH) recommended standard of 15000 people/EHP. A more recent survey provides a better indicator of front-line resourcing by calculating the number of 'functional-EHPs' that excludes managers and graduates completing their compulsory 1-year community service training (Agenbag and Lues 2009). Using this indicator created a new national average of nearly 46000 people per EHP, the Western Cape is still the only province achieving the NDoH standard. In February 2007 Urbington's functional EHP ratio was around 34000 people/EHP, more than double the NDoH recommendation. 6 years later this had improved to around 20400 people/EHP, but the NDoH standard remained some way off. Senior managers aspired to exceed the NDoH ratio as soon as possible but remained concerned that ratios distorted the picture in apartheid-planned cities like Urbington, where large numbers live in Townships but work elsewhere. Therefore they were exploring other variables relating to premises, activities and conditions to try to distribute their EHPs more effectively.

When asked about resources, Suburban EHPs considered them adequate but those in the Townships and Inner city rated them inadequate. However, EHPs across Urbington shared many concerns, particularly that there were just too few EHPs to cover such large areas. It could take weeks, even months to cover their areas, whilst focusing on one area meant others deteriorated and this was further compounded by population growth and the transient nature of many businesses. In response EHPs sometimes likened their roles to 'putting out fires' but they also sought to manage this by organising their work geographically where possible:

> … due to staff shortages you would wait a bit longer [to revisit a premises], so that the premises with 7, 14 or 30 days' notices expires and then you can visit all of them.
>
> (Township EHP 2)

Staff turnover created further pressures including the loss of knowledge and relationships and the additional support required by new and inexperienced EHPs. EHPs were also concerned about the loss of experienced Managers, including older White EHPs who had reached a glass ceiling and were retiring early without passing on their experience. A lack of resources was reducing core functions with the sampling of foods and drinking water suspended, pest control services limited to public advice only, staff training limited to law enforcement and projects ending long before they became sustainable. During fieldwork two regional offices were broken into with computers, printers and photocopiers stolen that were unlikely to be replaced for weeks. A few EHPs had access to work mobile phones but most used their own mobile phones (and credit) in the field. Without one it was very difficult to access certain areas like gated communities to investigate complaints. Township EHPs also commented on how the dilapidated conditions of their offices could undermine work:

> We say to the public they should do this and this and this and only when they come to your office they see the opposite of what you are preaching. So if the Council, because Health it's a core Department and it deals with the public and the clients visit them, they should set the trend … they should put their health staff in office which are lookable and which are kept clean all the times … .
>
> (Township EHP 1)

References

Agenbag, M. (2006) An analysis of progress made with the devolution of municipal health services, in South Africa Unpublished conference paper presented at SAIEH conference, eThekwini, South Africa, February 2006.

Agenbag, M. and J. Lues (2009) Resource management and environmental health service delivery regarding milk hygiene. *British Food Journal*, Vol. 111 (6) 539–553.

Beall, J., Crankshaw, O. and Parnell, S. (2002) *Uniting a Divided City: Governance and Social Exclusion in Johannesburg*. Earthscan Publications Ltd, London, UK.

Haynes, R. (2004) *Financing Environmental Health Services in South Africa*. Health Systems Trust, Durban, South Africa.

Kaplan, R. and P. Norton (1996) Using the balanced scorecard as a strategic management system. *Harvard Business Review*, Vol. 74 (1) 75–85.

6 The views of Urbington's EHPs

All EHPs acknowledged their work as law enforcers but considered their primary role to be more persuasive than punitive, with the law providing one means towards greater environmental health ends. Instead, EHPs saw themselves as educators and advisors, protecting people and improving lives:

> You have to help them because they need information … Some of them steal, some of them never went to school, so they need us, that's why we are always having this awareness of them. You know we actually convey the message, we are educating the client. It's good to see people improving in their lifestyle because you contribute …
>
> (Inner city EHP 2)

EHPs also saw themselves as promoting health and more sustainable lifestyles and preventing ill health by solving problems and responding to community needs. Recognised the inequalities in their areas, EHPs described their roles as trying to 'bridge the gaps' between developed and developing communities, a few even considering themselves as 'saviours' of the community, Inner city EHP 4 adding that without EHPs she didn't know where Urbington would go? Similarly, the Deputy Director of Environmental Health defined the role of his EHPs as the 'management of environmental health risks in the public interest' and EHPs often associated their work with delivering the *Constitution*.

But EHPs also recognised the need for more punitive approaches:

> …. when I arrived here I've realise that sometimes … I have to be an educator, but not all the cases will need … the education approach, some of the things they need you to be serious, they need you to be firm and say 'this is the law'.
>
> (Inner city EHP 1)

DOI: 10.1201/9781003009931-6

> ... you teach 1 day but tomorrow go three steps back ... as I'm teaching I think of all the inspectors that worked here before, I won't make any difference doing that. For me it's law enforcement, it's prosecution and then that's all that will make people change their mind. We are having people who are unemployed and they are here to make money, but they don't care about the law ... and they will NEVER even find out what the law says about those things.
>
> (Inner city EHP 3)

EHPs associated more punitive approaches with serious offences, uncooperative people and persistent offenders but they were also nervous that this could put them in a difficult position:

> ... when people see you as somebody who is there to help they are open, then we can help them, then it's better.
>
> (Suburban EHP 2)

> We must think about the poor and help them, but at the same time we don't want to be seen as the people sitting on top of other people'.
>
> (Inner city EHP 2)

Therefore EHPs generally resorted to punishment as the last resort and went to great lengths to persuade people that their interventions could benefit them and their customers. This ability to navigate the regulatory pyramid was evident when they were asked what makes a 'good EHP', the most common attributes included a flexible but firm approach, an ability to make decisions and work independently, good communication skills and being able to work with people from all walks of life. By respecting people, EHPs reflected that they were more likely to respect you and be cooperative. Other important values included professionalism (e.g. current knowledge, accountability), honesty, dedication and a passion for environmental health.

Some EHPs mentioned the poor environmental health they experienced during apartheid but only Inner city EHP 2 suggested it informed his EHP career choice when recalling an encounter with his school careers advisor:

> I grew up in a rural area and I've seen a lot of things that are being not addressed, and then coming to choosing a career ... this career it was actually addressing all those problems and I thought about

> that, the generation that is coming after me … now we need to do something and I said … give them education and save the community, that's my job …
>
> (Inner city EHP 2)

Careers advisors also steered two others towards environmental health but the rest stumbled across it following chance meetings with EHPs in previous jobs or via family and friends.

All acknowledged the broad and largely generalist nature of their work, even Suburban EHP 1 whose specialist pollution work as an Auxiliary EHP still required a great breadth of knowledge. A few likened themselves to 'jacks of all trades' but all combined their generalist district work with more specialist roles including communicable diseases surveillance, health promotion or managing complex housing cases. Most were satisfied with their jobs and when asked what they most liked about being an EHP they spoke of helping people, solving problems, promoting health, making a difference and characteristics of the job itself like working with different people, the variety of the work and not being office bound. All really valued their discretion and felt motivated by their work, particularly when this was acknowledged by the public.

Only two EHPs expressed dissatisfaction with their jobs, particularly around the lack of training opportunities at Urbington and working in the same areas for many years, though this was changing with EHPs beginning to rotate more between areas and regions. EHPs were frustrated when they failed to improve their areas or their work was not appreciated by managers and the public but pay and benefits were the most common cause of dissatisfaction. For some Urbington's pay and benefits were an improvement over previous jobs but all were concerned that their car allowances did not cover their costs, particularly following increases in petrol prices. Inner city EHP 3 also linked her pay concerns to her job title:

> … what happens is we are still regarded as Environmental Health Officers … and an OFFICER is someone who's got matric (i.e. a high school graduate) and … a qualification to work as let's say a Metro. Police Officer they just want a matric and a driver's licence, and an administrative officer, they want your matric … So we are more qualified, we've got degrees, we've got Diplomas … So that's why the Officer part of it has an impact on our salary, because we get paid as Officers, we don't get paid as qualified practitioners …
>
> (Inner city EHP 3)

To establish this a group of EHPs had collected job adverts for 'Officer' posts from across South Africa and identified lesser qualified posts in the same salary range as Urbington EHPs. Further, Inner city EHP 3 explained that the law usually referred to inspectors, not practitioners, but momentum was building towards legal recognition of the title 'Environmental Health Practitioner' and this was finally achieved in 2013 as acknowledged by the President of the SAIEH:

> Environmental Health Officer, to us, meant a professional who would just be looking at enforcement. An Environmental Health Practitioner is a person who should scientifically make a proper assessment of environmental health conditions and institute appropriate mitigating measures that would sustainably address environmental conditions. This is a true description of what we are trained to do, as professionals practising within the field of environmental health.
>
> (Chaka 2013)

Another concern about pay was the difference between Urbington and other cities where EHPs knew they could earn around R2500 (~$350) more, plus higher car allowances and merit increases in salary. Two openly admitted they would move jobs if a suitable post came up. A few even nicknamed their employer the 'University of Urbington for Environmental Health Practitioners", training South Africa's newly qualified EHPs before they moved to better-paid jobs elsewhere. Indeed, high levels of staff turnover were common across Urbington whilst remaining EHPs were concerned about its knock-on effects, particularly how areas deteriorated with the loss of local knowledge and relationships.

Some EHPs were also concerned that their colleagues, particularly younger EHPs, were too preoccupied with money:

> You are not just here to earn a salary, you are here to make a difference
>
> (Suburban EHP 3)

Others thought EHPs should be focused more on how to improve their areas before asking about money but there was general agreement that they would be happier if Urbington looked after them better. Opinions varied about the differences in pay between EHPs themselves. Most EHPs did not think their managers were paid fairly, Township EHP 1 commenting that if they were they wouldn't keep resigning as her own

Operations Manager had just done so. On-going changes to pay scales also created tensions, with some newly appointed EHPs earning more than experienced EHPs, the latter then feeling less motivated to support their inexperienced but better-paid colleagues. All EHPs were Union members and fieldwork coincided with public sector strikes across South Africa. Though Urbington was unaffected at this time, EHPs were also working with the unions and their employer to address their pay concerns and there was some optimism of a resolution. Most EHPs would also recommend a career as an EHP to a suitably qualified friend, though a few would add a 'friendly warning' about the realities of working at Urbington, particularly the pressure on EHPs to meet targets.

The views of EHPs as professionals

All EHPs considered themselves professionals, not just due to their mandatory HPCSA registration to practice environmental health but in other aspects like working hard, being up to date, committed and accountable. EHPs were also united by a concern that their profession was under threat and that the HPCSA could be doing much more for them beyond setting standards and regulating their practice. Indeed, some complained that they got nothing from HPCSA membership, which they paid themselves, and only heard from the HPCSA when fees were due. Some suspected this was because the HPCSA concentrated on nurses and doctors but there were signs this was changing, including the introduction of Continuous Professional Development (CPD) for EHPs. But this was accompanied by concerns about the lack of CPD courses available and whether de-registration, if you don't meet new CPD requirements, was wise during a nationwide shortage of EHPs.

Before democracy, there were separate Black and White professional associations for EHPs but these combined to create the South African Institute of Environmental Health (SAIEH) in 1995. The SAIEH was working to promote the science and practice of environmental health and organised national and international conferences that enabled EHPs to share knowledge and experience and influence policy, though only a few Urbington EHPs knew of this work.

Though environmental health cuts across legislation and Urbington strategy (see Chapters 4 and 5), EHPs described poor levels of environmental health awareness within Urbington itself. The Director of the Department of Health claimed Environmental Health was no longer a Cinderella service and wanted EHPs to assert themselves and claim their rightful role in delivering health services alongside doctors and nurses, but EHPs were less convinced. Amongst their concerns were the loss of some traditional functions to other Departments with their own inspectors (e.g. Environmental Management) and threats from new posts, including EH Assistants:

> ... it looks like our qualifications are going to fade because they are going to look at any person who only have Matric to do environmental health ... they are bringing in the assistant which will be 2 years and ... now they wanted to come up with the Police in our job, with only Matric ... they will have trainings even though you don't have a diploma or a degree or whatever in environmental health that will actually be the health inspector.
>
> (Inner city EHP 3)

> ... already Urbington Police wants to test diesels, Environmental Management wants to take over air pollution and water pollution .... originally your Act states that these are the functions of the Health Department within local government, but now if we don't take it seriously and we don't do it, it will go. If environmental health opened their eyes what will be left, just vacant stands and barking dogs and things like that, which you can give to an assistant to do. So if they don't take their work seriously, this is the way they will go. I think it's just they must take their work seriously. And our management as well, they must, I think they just need proper guidance on what is environmental health.
>
> (Inner city EH Manager)

EHPs were also concerned that their service remained a Cinderella:

> The profession is not marketed. We are not known and we are just being taken as the people who just pass Matric and then are given a job to go and CHECK waste water in the street, go and CHECK building rubble on the sidewalk. [Senior managers] don't know what it involves, how important it is.
>
> (Suburban EHP 3)

> [Director – Department of Health], you know what you are doing with us, if you take a doctor and you ask him to make beds in the hospital, that's what you are doing with the EHO. We concentrate 50% on vacant stands, on grass cutting, on things like that you can

> give to somebody who's just come out of school to do it. So I compare it to a doctor who's making beds in hospital. You don't use the EHO for what he was trained, you demote him ... We could rather spend our time and make sure that food premises complies, that we take food samples which we don't do ... Look at the air emissions, what's happening in the factories, those kind of things. We have studied for that. But I mean grass cutting, if the grass is higher than this I'll still debate it, is it really a health nuisance? Grass is higher than me ... is somebody going to die of long grass? But they can die because of a food factory. I've got a food factory here, it's horrifying. They are manufacturing sauces in a panel beating shop, that's our work, to go and sniff out those places. Grass cutting, I mean, really, give it to the Parks department.
>
> (Inner city EH Manager)

In response, EHPs thought the HPCSA, the SAIEH and the National Director of Environmental Health should be doing more to protect and promote EHPs. By 2013 progress was being made, including an EHP (and former SAIEH secretary) appointed National Director of EH and the SAIEH and NDoH contributing to South Africa's first National Environmental Health Policy (NDoH 2013) and establishing a National Environmental Health Forum (Chaka, 2013). In 2012 the Professional Board for EHPs of the HPCSA also started publishing a bi-annual newsletter that included educational articles for EHPs like Ngqulunga (2013) on the need for persuasion and punishment.

The personal safety of EHPs

One inner city EHP commented that 'anything' can happen when you're working but all EHPs agreed their safety could be managed much better. All had been threatened in their work and being obstructed, warned to 'go away' or having their paperwork ripped up was not uncommon. Some had been locked in rooms, threatened at gunpoint or had their cars vandalized. Two EHPs recalled stories of past murders of EHPs though I was unable to verify these accounts. One involved an EHP inspecting a warehouse containing unlabelled packages; the warehouse operators later killed the EHP because they thought he knew they contained illegal drugs. Another murder followed the removal of telephone containers from street corners used by hawkers and others for trading. In response local people were unhappy, mistakenly accused the local EHP of organising their removal and killed him. A third case involved a Township EHP who served a spot fine on someone and was later killed for doing so. As one EHP commented:

> ... at the end of the day, it's our life involved
>
> (Inner city EHP 2)

Inner city EHP 4 thought being a White female EHP put her at greater risk and that might explain why so few White women now became EHPs. Amongst male EHPs there was rarely any 'macho safety culture' observed, indeed all were very open about the dangers of their work and admitted they were sometimes frightened, particularly when encountering the factors in Table 6.1.

Some areas had reputations for danger and included those considered 'no go' zones by EHPs that could only be visited with police backup. Work in remote and/or inaccessible areas was potentially more hazardous but EHPs were always on the lookout for people, snakes or other hazards (e.g. unstable river banks) for example. They also took great care working in certain places like bad buildings and informal settlements, conscious that tensions (e.g. fear of eviction) could get them into trouble unless they worked with

Table 6.1 Factors EHPs associated with increased personal safety risks

Areas and premises
Areas with reputations for violence
Remote and inaccessible areas
Informal settlements, bad buildings, nightclubs, prisons
Premises hosting 'illegal' activities (e.g. shebeens, hijacked buildings, workshops)
Characteristics of people
Uncooperative people
Tasks
Working alone
Entering buildings
Working higher up the regulatory pyramid (e.g. spot fines)
Blitz work

local people from the outset. EHPs were very wary of premises that might be hosting illegal activities, particularly when making the decision to enter them. For example, one EHP recalled the shouts of a caretaker when visiting some flats for what she thought was a routine inspection:

> Leave! If you don't leave, you see this street? [Caretaker takes out his gun and points it at EHP 1] I'm going to blow your head and your brain will be scattered there.
>
> (Township EHP 1 imitating caretaker)

EHPs were wary that encounters with uncooperative people could turn violent when, if possible, they would leave immediately and call for backup. Sometimes this didn't work and EHPs recalled being locked in rooms or thrown out of premises. Inner city EHP 2 admitted he was still very shaken months after being thrown out of a building and understood why EHPs were reluctant to return to areas they had been warned to stay away from.

Great care was always taken when entering buildings, particularly those in more violent areas or away from busy streets. More punitive sanctions like the spot fine also increased the risk because EHPs didn't know how people would react. Blitz work targeting offending hotspots exposed EHPs to greater danger and sometimes required them to wear bullet-proof vests but during this work, they were also better prepared and accompanied by the Police and others.

Some safety controls were in place. Though EHPs sometimes did not routinely notify colleagues of their whereabouts they had a very good working knowledge of their areas. Those areas/premises considered higher risk were only visited in numbers and/or with the Police. When encountering unforeseen dangers EHPs left immediately and returned with the appropriate backup. Most felt supported by their managers here, though some described a perception amongst managers that EHPs brought trouble upon themselves and should therefore deal with it themselves. EHPs were offered counselling after incidents and could request Police support anytime but in reality, this could take hours because the Police themselves were short-staffed. Some knew their local police well but others recalled dialling 10111 [the South African emergency services number] in desperation when in trouble. A new safety task team for Urbington was being established to investigate these problems and controls further, with one EHP per region, and EHPs were hopeful their concerns would be taken more seriously in the future by all those involved (e.g. managers, Police, Unions).

The EHP as Peace Officer

All EHPs were Peace Officers authorized by Urbington to enforce the law, though new starter Suburban EHP 1 had not yet completed the Urbington law enforcement qualifying course and therefore had to call on his colleagues when sanctions were required. EHPs recalled studying law enforcement at University but some thought more was needed here:

> … I think we must get into the Universities, they must really look at their curriculums, make it more what the EHPs are doing on the job. Because I think sometimes Universities create a picture for them that is not a reality, and they get the shock of their life when they start and see they must prosecute this person [laughs] and all those things, like that nuisance building here and you throw a junior EHP in there and he won't know where to start, what must he do, and the notices that they must write. I think that the lecturers are a bit out of touch with reality, with the conditions that we have here … Sometimes I think they must get real environmental health, people that's working to come and input into the lectures and say 'this is the kinds of things that you get, this is how you must solve it' …
>
> (Inner city EH Manager)

EHPs recalled Urbington's own week-long law enforcement course and cited benefits including an awareness of regulatory procedures and the courts, the legal rights of offenders and their own responsibilities and accountabilities as Peace Officers and local government officials. Course ring binders full of legislation and standards sat beside their desks for quick reference and all valued the yearly refresher courses on topics including new legislation, legal loopholes, writing fines correctly, how to build a case and how to manage people. Managers were concerned that it might take up to 6 months to get new EHPs trained as Peace Officers but the experience was also vital:

> … if I can get them trained tomorrow they have got a piece of paper but they haven't got the experience and the knowhow and the confidence to go out and do it.
>
> (Suburban EH Manager)

For managers, this also reinforced the need for the ongoing training and close supervision of all EHPs. The HPCSA had also commissioned an article in one of its newsletters reinforcing why carrots and sticks were both necessary:

> There is always an expectation that all Environmental Health Practitioners constantly engage in the health education of the consumers of our services, particularly the general public. Our task is to educate, educate, and educate .... It does not, however, work single handedly. It is accompanied by awareness campaigns involving health promotion, roadshows, street drama, etc. The question still remains as to what needs to happen when all of these fail to achieve the desired results? ... It is expected that when all else has failed, people who violate the right of the public to a safe and clean environment should be brought to book.
>
> (Ngqulunga 2013:8)

The general views of EHPs on law enforcement

EHPs were largely complementary about the law. Those with apartheid-era experience reflected how things had changed from highly prescriptive laws, strictly enforced, to more accommodating laws and more flexible approaches, but this transition was a great challenge that required a far broader knowledge and a change in ethos:

> [now you need] to know WHY are you asking for something, not just 'the law says ...'
>
> (Suburban EH Manager)

When asked, half of EHPs still preferred laws with precise definitions of standards when compared to those requiring individual interpretation, but in the field, EHPs valued the greater discretion and flexibility of the new legal framework, particularly when regulating complex cases. However, some were concerned about what their colleagues, particularly those with limited experience, were doing:

> ... to one inspector something can be bad, and the other one think 'ag, he can still get away with that'
>
> (Inner city EHP 4)

Further, EHPs didn't recognise many of the legal uncertainties identified in Chapter 5. Instead, as Hutter (1988) found, these were framed as 'problems to be solved' with EHPs more concerned about the practicalities of their work, including the 'Biblical' volumes of law as one senior manager described them. EHPs constantly referred to the law and there was general agreement that some laws, particularly Urbington's *Public Health By-laws*, were easier to understand than others. After 5 years at Urbington Suburban EHP 2 had memorized most of the laws he used daily but he regularly encountered cases where the law and possible actions were not clear and would call his colleagues or managers for assistance. EHPs were also concerned about the lack of guidance in certain areas, particularly for closing premises, but this was being addressed by senior managers.

Some of the greatest legal challenges EHPs identified included regulating the informal sector, as covered in Chapter 7, and working on long-term cases. For example, housing cases that required the tracing of building owners could take years whilst the tenants, often living in very poor conditions, remained silent for fear of eviction. EHPs were also concerned that their effectiveness was limited because they didn't routinely work evenings or weekends. Following up the next working day might only result in a warning at most because the evidence had disappeared. Requirements for permits or licenses also presented many challenges with significant backlogs in applications across Urbington, whilst EHPs constantly discovered new premises. The payment of fees for some permits or licenses also raised expectations, this transaction creating what some EHPs called an 'unwritten promise' of a permit. Standards could also drop after receipt of a permit because some businesses considered it their right after payment, not a standard that had to be earned and maintained. Relations between EHPs and applicants could also become strained when multiple licenses were required, particularly in settings like nightclubs which had a reputation for assuming that once they received a liquor license they were free to open.

Law enforcement training included the management of legal loopholes but some continued to be exploited and took up considerable time. For example, Urbington's *Public Health By-laws* required urgent review because its requirements for accommodation establishments placed the duty to apply for a permit on the person who conducts the establishment, not the owner, and was being exploited by armed criminal gangs who hijack buildings:

> Let's say I hijacked your house and chased you away. I go and apply for a permit. I've cleaned my house, everything is nice and clean, the health inspector comes and says 'oh, this is fantastic, let me process your permit because everything is in order'.

> He gives me a permit. They [EHPs] don't enquire whether I'm the owner, the fact that you applied HERE it's like you have a right to occupy the house ... I get a permit ... I rent it to different people and they will have the confidence that this person is the owner of the house because he has given him or her documentation.
>
> (Inner city EHP 1)

Until the law was changed, to avoid legitimizing criminal activity EHPs were spending more time checking property ownership and were working closely with Urbington's Legal Department on suspect cases.

Urbington published and distributed environmental health information and advice in all eleven South African languages and others (e.g. Mandarin). English was the official language of regulation but most visits were conducted in other South African languages and EHPs were at a disadvantage if they did not speak them. No translation services were available but EHPs supported each other with translation or enlisted the help of the regulated and others (e.g. friends and relatives) during visits. Non-South African languages were more common in some areas (e.g. inner city) but EHPs were happy working around this provided people were cooperative. More challenging was that all paperwork was in English and therefore EHPs sometimes spent considerable time writing in English and, where possible, the language of the regulated too. Further, if they suspected a person might be illiterate EHPs spent more time explaining and demonstrating what they wanted before their next visit.

The uncertain criminality of environmental health offending

Crime was a constant discussion topic. As one EHP put it, South Africa has too many crimes and environmental health just wasn't a priority. Another recalled a nightclub noise complaint whose owner responded along similar lines:

> 'Ok, what is wrong? Did you see, there was this woman who was killed the other day, and you are telling me I'm making noise!'
>
> (Suburban Operations Manager)

Reflecting on the lawlessness he'd witnessed in Urbington, one Suburban EH Manager described people who dump rubbish on the pavement because 'for the last 5 years nothing has happened to me' and the Mayor himself highlighted this during a public speech:

> It is unfortunate that people who complain about so-called 'big crimes' such as robbery or murder often have no compunction to commit 'small crimes' such as traffic offences, illegal dumping, engaging with acts that lead to pollution and some even fail to pay rates or taxes. I urge you to observe the speed limit, pay your traffic fines, do not dump your rubbish on open sites, and pay for the consumption of services such as water and electricity. Repent and serve the community.
>
> (Mayor of Urbington)

EHPs varied in their views of the criminality of offences. Suburban EHP 1 thought all offences were criminal because, regardless of the offence, contraventions of the law were an offence for which you could be prosecuted. Others thought only some offences were criminal because most of their work was about promoting health, not punishing people. Inner city EHP F1A for example stressed that EHPs only use criminal procedures when people behave in criminal ways, but he was hesitant to criminalise people because environmental health was a Department of Health function. EHPs were very reluctant to describe EH offenders as criminals. There were no typical offenders because each was unique but EHPs instead categorised most people as Kagan and Scholz's (1984) *organisationally incompetent*, i.e. they broke the law because they lacked knowledge and in response needed education and persuasion that 'this is wrong, this is what you need to do' as EHPs often put it.

Technical offences were less likely to be considered criminal compared to those where motive or intent was involved such as intentional acts (e.g. obstructing EHPs, using adulterated cooking oils, timing offending outside EHP working hours) or omissions (e.g. failure to provide ID, trading without a permit), persistent offending and where EHPs suspected profits were being put before people. More 'criminal' offences also included serious offences (e.g. illegal dumping, building hijacking, food adulteration) and those affecting the public, particularly children and the poor. Occupation was also a factor, with nightclub and bad building owners more commonly associated with uncooperative behaviours and profit motives. The business size was sometimes a factor, with larger businesses that 'should know better' when compared to smaller businesses and the informal sector needed support but could also be evasive and dishonest.

EHPs' views of sanctions and the criminal courts

The spot fine was the most common sanction used by EHPs, usually in response to intentional and/or persistent acts of non-compliance and when other approaches had failed. Serious offences and those affecting members of the public were also more likely to result in a spot fine, particularly during blitz work targeting offending hot spots. When asked about the spot fine EHPs thought they were needed to punish offenders, deter future offending and encourage compliance by making people aware that they had legal powers. EHPs also valued the threat of criminal sanctions when using persuasive approaches, for example warning people that an offence could result in an R500 fine or that spot fines are the same as those used by the Police force, where non-payment leads to a Court summons and possible prison sentence. EHPs also sought to persuade by warning of their futility:

> So you show them that the R1000 that I'm going to fine you, you would have repainted the wall as I requested, you would have bought paints and painted the walls. So now you're going to give me that R1000 because you don't want to paint the wall and I've got to come back AGAIN next month to give you another R1000 because you won't have money to buy the paints again.
>
> (Inner city EHP 3)

But EHPs also had concerns about spot fines. They questioned their effectiveness unless combined with education and were wary they could put their safety at risk, but all provided examples of persistent offenders who simply paid their fines and continued offending. A procedure existed to prevent this, in theory, by enabling EHPs to force persistent offenders to court without the option of paying fines but its effectiveness was questioned as described further below. The maximum sanctions were another concern, some EHPs defended this because, like the traffic fines on which they were based, spot fines were standard for all offenders regardless of socioeconomic status. However, EHPs were well aware of the consequences of this:

> … in the [formal] day care centre a fine is R1000 and then in the informal day care, to them R1000 ($140 in 2007) is almost the whole profit which they are generating a month.
>
> (Township EHP 1)

Whilst those on low incomes complained that they were unable to pay spot fines, for others they were too low to be effective or penalized others unfairly. For example, nightclubs had a reputation for temporarily increasing their admission fees to cover the costs of spot fines. But the greatest concern was that EHPs did not know what happened after a spot fine had been served and this could undermine their authority:

> We can work so hard, 100% for everything, but did we really make an impact? … We've written how many, 200 and something fines, so what? We don't get the money back … does it really make an impact? How many of the fines were paid, we don't know … we're not even in control, it goes to Urbington Police's hands.
>
> (Inner city EH Manager)

> … last year I remember I gave this guy a fine and when 6 months later I saw him … he didn't pay the fine, he just took it. I gave him another fine because it was keeping on again, giving notices and he just took the fine and put it away in a draw. And I wondered who follows up these fines if they are not paid? … he doesn't respect me when I go, because he knows that whatever I say he's not going to do it, because there is no mechanism to force him to do it.
>
> (Inner city EHP 1)

Some Inner city EHPs, therefore, nicknamed the spot fine a "dead snake" and had completed their own investigations. They found many were dismissed in court because of errors, though EHPs were adamant that there were none and suspected corruption as explored further below. One Suburban Operations Manager had served 20 spot fines recently but on visiting court discovered only one had been paid, some were not recorded and vital control documents she had completed were missing. Public Prosecutors were complaining about mistakes in paperwork and this manager illustrated this with some recent examples on her desk that would be rejected by the courts and included no time recorded, lost carbon copies, no ID number, no initials and scratches on carbon copies. Senior managers also acknowledged these problems and were addressing them via supervision and ongoing law enforcement refresher training. EHPs were also wary that some informal businesses seemed to be taking advantage of this situation, something she likened to an 'on-going circle' that could only change through persuasion, not punishment:

> ... we've got a lot of people with premises ... in permanent buildings. THOSE people pay their fines, because they know tomorrow we'll get them there again ... they've got nowhere to go. Now you've got ... the hawkers, the informal traders, the people who are like selling food on the corner, you [serve a spot fine], they give you the wrong names, the wrong addresses, they haven't got ID books .... tomorrow they're back on the same piece of land because they've given you the wrong information and they don't pay it. So [spot fines] don't help.
>
> (Inner city EHP 4)

EHPs were therefore working with the Police following up unpaid fines and persistent offending with warrants of arrest but they remained sceptical of the local criminal justice system because it was already overloaded and environmental health matters just weren't a priority.

EHPs also feared some public prosecutors might be corrupt. Alongside many unsolved cases of perfectly completed spot fines disappearing, one EHP recalled a case involving a persistent offender who'd been subpoenaed by the Court but the case was delayed. They went to the public prosecutor to find out why and discovered that the case had been withdrawn. The anonymous EHP suspected corruption, with the business owner probably approaching the prosecutor directly to avoid a court appearance. The same EHP recalled a similar case where a Ward Councillor intervened at the request of a business owner by contacting the prosecutor who then withdrew the case because, in the politician's opinion, the owner now complied. The EHP was furious having spent weeks on the case only to be told by the prosecutor, who had not worked on the case before, that it had been withdrawn and senior managers were reluctant to follow it up. In response the EHP commented:

> You prosecute because you're serious ... [but] no one will back you.
>
> (Anonymous EHP)

The incorporation of fines within performance targets was another common concern of EHPs, as reviewed in Chapter 5. Though it was not difficult to meet or exceed such monthly targets EHPs were more concerned about being pressured in this way, particularly given their preventive ethos and the potential for sanctions to harm their relations with businesses and the public. Greater emphasis on alternative sanctions like prohibitions or the seizing of equipment was also considered important and further guidance was under development by managers and Urbington's lawyers.

Most EHPs had little or no court experience because prosecutions were usually organised by managers and public prosecutors though EHPs were well aware of the role of the courts and their own legal accountabilities. They also knew offenders did not like going to court, particularly when fines could be increased, but they also feared the courts would be lenient on offenders as one EHP recalled from a recent case:

> 'I was supposed to be dealing with criminals, now I'm busy with a person who is DUMPING [waste]! What is dumping after all?'
>
> (Words of magistrate to Township EHP 1)

Another case had become folklore across Urbington's EHPs. A suburban manager had forced a persistent noise case to court but the magistrate opened with the following words:

> 'What is this? [Magistrate taps on his desk]
>
> What is this? [Magistrate turns to Suburban Operations Manager]
>
> Lady, who brought you? What's your name? You are just wasting my time in court!' …
>
> [Magistrate turns to addresses defendant]
>
> If this woman calls, I will just throw this out, because you are wasting the Court's time'
>
> (Suburban Operations Manager)

The manager reported this case to senior managers but had some sympathy with the magistrate:

> … we have the crime problem in the media, it is a priority, you know, a shortage of police … you hear all the time on the radio and there is an outcry, you know, people of South Africa … crime and all that. And when you go to these courts the Magistrates know nothing about all these environmental laws. They know nothing. And it is sad that you have to take in the same court where a person who is being tried for rape, murder, burglary, car jacking. You have

> to take the same court as somebody who is playing a radio, who is making people not sleep ... the reality is that if I was a magistrate this same place where you see the same people in chains where they are being brought up from the cells in the same court! With it, the scale really, you can't even comprehend
>
> (Suburban Operations Manager)

Another EHP's review of the previous year's law enforcement statistics reconfirmed the spot fine as 'dead snake' argument:

> Last year more than 1200 environmental health cases were recorded, ONLY 12 of them in Urbington were taken to court and successfully prosecuted. The rest, NO! So that's what, 1200 versus 12, you can see that something needs to be done, because we are telling them to comply but nothing happens.
>
> (Suburban EHP 2)

In response EHPs agreed that the judiciary needed training in environmental health law. Some also argued for more dedicated courts for EH but managers reported that previous trials of this in Urbington had failed due to insufficient numbers of cases. The Department of Health had also recently secured the assistance of a dedicated public prosecutor for EH cases but following a series of missed meetings EHPs were not full of confidence. More encouraging was developing collaborative approaches with other Urbington law enforcement organisations, particularly the pooling of experience to develop guidance on closing premises and confiscating equipment.

References

Chaka, J. (2013) SAIEH President's message. *South African Institute of Environmental Health Website*. Accessed on 1 February 2014 via: www.saieh.co.za.

Hutter, B. (1988) *The Reasonable Arm of the Law? The Law Enforcement Procedures of Environmental Health Officers*. Clarendon Press, Oxford, UK.

Kagan, R. and J. Scholz (1984) The 'criminology of the corporation' and regulatory enforcement strategies, in K. Hawkins and J. Thomas (Eds.) *Enforcing Regulation*. Kluwer-Nijhoff, Boston, USA.

NDoH National Department of Health (2013) *National Environmental Health Policy*. Government Gazette No. 37112 4 December 2013.

Ngqulunga, G. (2013) When all else fails, in *Environmental Health News 2013/14* Newsletter of the Professional Board for Environmental Health Practitioners of the HPCSA, Pretoria, South Africa.

7 Relations with the regulated

Typologies of the regulated and complainants

All EHPs described their areas as a combination of 'undeveloped' or 'developing' alongside 'developed' areas where the ability to work with people from so many different backgrounds was critical. Their views of the regulated were frequently shaped by moral considerations and sometimes EHPs, like Lipsky's (1980) street-level bureaucrats, stereotyped people in an attempt to conclude cases as quickly as possible. But revisiting Kagan and Scholz's (1984) images of the regulated in Table 3.1 I also discovered a complex world where such people's processing was more difficult. This was because EHPs viewed the regulated as mostly *organisationally incompetent* or Beall et al.'s (2000) *victims*; i.e. well-intentioned and inclined towards compliance but offend due to their lack of knowledge and/or their socio-economic circumstances. One EHP framed this using Maslow's hierarchy of needs:

> If you haven't met the first level, how can you go to the third level and complain about the services? You cannot take money which you want to buy bread or buy clothes ... and have to make a telephone call to lodge a complaint.
>
> (Township EHP 1)

Here EHPs typically responded using persuasive approaches to support the regulated that reflected their own motivations to help people and sometimes, they admitted, their own failures to educate and inform the public. EHPs also identified small numbers of Kagan and Scholz's (1984) *amoral calculators* or Beall et al.'s *villains* motivated by profits and whose offending results from economic calculation. Typical examples included illegal dumpers, building hijackers,

DOI: 10.1201/9781003009931-7

the owners of bad buildings and some nightclubs. In response, EHPs were likely to use more punitive approaches.

EHPs also identified a few of Kagan and Scholz's (1984) *political citizens* who are inclined to comply with the law but offend due to principled disagreement with laws they consider authoritarian or unreasonable. For example, the owner of an overcrowded township crèche accused Urbington of delays with her permit before describing far worse conditions in other crèches and accusing EHPs of deliberately targeting her. The owner was disappointed with Urbington and concerned about what parents of the children about to be excluded would think of their EHPs. In response Township EHP 1 acted like Kagan and Scholz's politician, recognising the high standards of the crèche but insisting that the crèche must reduce its maximum number of children by 10. EHPs also labelled some as Beall et al.'s (2000) *fixers*, helping to improve local services and environmental health conditions. These included those working with EHPs on projects or promotional activities like schools, businesses, charities, community groups and individuals.

Ayres and Braithwaite's (1992) multiple selves argument also applied to some people occupying more than one typology at the same time and/or moving between typologies. For example, Suburban EHP 3 considered the residents of a new housing area to be *victims*, having previously lived in an informal settlement, and villains, following the emergence of new shacks and abandoned vehicles there. EHPs often viewed street hawkers as both *victims* and *villains*:

> [P]eople just SIT wherever they want to sit, SELL wherever they want to sell because they want to make money. Whether that place is prohibited for selling or not it's none of their business. People prefer to sit on the road and do it because they want to make money … And at the end of the day we understand they are trying to make money, but make money in a way that you will never be against the law but they don't understand that. It's so, so difficult because today we get a Police van taking their goods, removing them, tomorrow you have extra people sitting there. So it's actually not taking us anywhere.
>
> (Inner city EHP 3)

Township EHP 1 recalled a complaint investigation in which the owner of a cottage factory began as the *villain*, behaving very aggressively towards her with the help of his large dog, but also the *victim* who was reluctant to move to a nearby factory area for fear of losing his livelihood and house. EHP 1 applied increasingly punitive approaches during her first revisit, including the threat of a spot fine but the owner later transformed into a *fixer* when he moved and his business started to grow in its new factory site. He also apologised to Township EHP 1, even praising her as a blessing in disguise.

A complainant's motivations were also an important factor in determining their actions and generally followed four of Hutter's (1988) categories as summarised in Table 7.1. These were not set in stone and the 'multiple selves' argument also applies here, for example when investigating residential noise complaints EHPs reserved judgment on whether the complaint could be genuine or a mix of intolerance, psychological factors or a possible dispute until further investigation. Unlike Hutter (1988), no evidence of financially motivated complainants (e.g. compensation seekers) was found but there is no reason why this category shouldn't apply in Urbington too.

Table 7.1 Four main categories of Urbington complainant

Complainant type (From Hutter 1988)	*Fieldwork examples and potential responses of EHPs*
Genuine	For EHPs these included serious cases with immediate potential health risks (e.g. allegations of food poisoning, pest infestations, sanitation problems) and resulted in rapid and often more punitive responses.
Intolerant	These included some noise complaints, particularly from retired people in suburban areas. EHPs investigated them and treated these complaints seriously but used mainly persuasive approaches to try to resolve them.
Psychological	These included complaints from people who EHPs suspected may be lonely, unhappy and/or had nothing better to do. In response EHPs usually visited the alleged offender and used persuasive approaches whilst reassuring the complainant that the issues were being resolved.
Neighbourhood disputes	Where possible EHPs tried to avoid getting involved in these cases, though such disputes sometimes weren't obvious at the early stages of investigation and often related to alleged nuisances (e.g. noise, litter). Where possible EHPs tried to smooth over the situation using informal approaches and tried to avoid further involvement.

Other characteristics of complainants were also important to EHPs, particularly socio-economic factors. EHPs frequently differentiated between more affluent and vocal complainers and those from poorer backgrounds who complain rarely, if at all. In response, EHPs were careful when handling the former, particularly if lawyers were involved, whilst trying to improve the access of the poor to the complaints process, for example by spending more time in poorer areas or attending local meetings in person. EHPs knew that ongoing complaints made people angry but they also felt that sometimes people didn't appreciate their complexity. The involvement of local politicians increased the speed of complaints investigations and EHPs were careful to keep their managers constantly updated and were on the lookout for other motives, particularly around election times. The preferences of complainants were also considered, but sometimes EHPs decided otherwise. For example, during one smoke investigation, the complainant was adamant that Suburban EHP 1 should contact the property owner directly and not the builders next door because she feared they would just laugh and know she complained. But Suburban EHP 1 disagreed and visited the builders straight afterwards who were very apologetic and switched to using smokeless stoves soon afterwards.

The severity and persistence of offending

The severity of actual offences or the potential for serious offences were important influences and EHPs constantly weighed up their potential to cause harm, the likelihood of harm occurring and the persons likely to be harmed. For EHPs severe offences included illegal dumping, bad buildings, food adulteration and poisoning and offences affecting the public, particularly children and the poor. In response, EHPs were more likely to use more immediate and more punitive approaches like the spot fine, though all options were considered. Persistent offending was also more likely to result in punishment, sometimes regardless of whether the actual offence was serious. Its association with criminality through intentional acts/omissions was important here, Inner city EHP 2 recalled a shop that continued to sell rotten foods to the poor at a discounted price where he served a spot fine immediately on his first revisit and continued to monitor the business closely.

The character of the regulated

For EHPs this comprised many factors including the behaviour of the regulated, the physical conditions observed and the history of compliance. As Hutter (1988) found, most encounters were cooperative and EHPs were more likely to use more punitive approaches in response to uncooperative and aggressive behaviours. This was partly informed by their training, where EHPs recalled being taught that they should always listen first but people could be uncooperative for many reasons, sometimes as a deliberate tactic. The initial reception provided important behavioural cues and all EHPs knew that their own approaches could elicit greater cooperation and spoke of the importance of 'getting to know people', 'being in touch with them' or 'being human'. EHPs were also well aware of historical attitudes towards them characterised by distrust, fear and sometimes hostility and that building relationships took time. Suburban EHP 3 summarised his own approach as one of empathising with people, sitting down next to them and discussing their problems; only after this could you start talking about environmental health matters.

EHPs also recognised the importance of respecting people and hoped that when they knew more about their roles, beyond crude 'health inspector' stereotypes, they would become more cooperative:

> … people are more afraid of what is it that you are going to do and your attitude towards them.
>
> (Inner city EHP 3)

Therefore clear introductions, particularly when meeting people for the first time, made cooperation more likely and at the start of every visit EHPs took time to explain who they were and what they would be doing. Titles were also important and despite many good reasons to be recognised as EHPs, they usually introduced themselves as 'health inspectors' as the public usually referred to them.

EHPs themselves could also help create uncooperative encounters. For example, Suburban EHP 3 imitated his colleague by dropping a pen on the floor, kicking it around whilst holding a clipboard and repeating 'what is this, what is this' in an aggressive manner. He put this down to his colleague's lack of experience but was also concerned that some EHPs did not appreciate the danger they could be putting themselves in. Suburban EHP 1 was the only inexperienced EHP observed and whilst some EHPs were more formal than others in their work, none were observed being as aggressive as Suburban EHP 3 recalled. Interestingly, Inner city EHP 2 was unsure why nearly half the people he encounters did not respect him but he thought his (young) age might be a factor as the 'old guys' he worked with were generally well respected by the public.

Two observed visits were quite hostile, with EHPs being shouted at, one echoing the comment that:

> … some people, even if you approach them in a good manner they always have a bad attitude towards you.
>
> (Inner city EHP 1)

In response, the two EHPs involved remained calm but became noticeably less flexible. However, they also gave people the opportunity to complain directly to their managers and provided all their contact details, an exchange which helped further calm things.

Friendly and cooperative behaviours could result in persuasive approaches even in potentially serious cases. For example, during food premises revisit a cooperative manager explained to Inner city EHP 4 that the problems identified in her last visit had not yet been fixed because the restaurant was closing soon and relocating but the staff had not yet been told. The EHP also found evidence of smoking in the kitchen but she accepted the manager's reassurances that all problems would be rectified immediately or by her next revisit. The physical conditions of the premises also provided EHPs with indicators about the regulated. Those running premises that appeared clean and well-maintained suggested a conscientious character that advantaged large, formal businesses:

> Your formal food premises, your [National company], when we went there, those are the ones that can comply because they are huge … I tell them without blinking: 'I want a double wash hand basin, I want wash up facilities', because the structure is built exactly for that.
>
> (Township EHP 2)

At the other extreme small and informal businesses were inherently disadvantaged but EHPs knew this and used their discretion to develop pragmatic ways towards compliance. EHPs expected a more hostile reception when visiting informal premises but were often impressed at how standards were maintained there, for example, a clean and well-maintained shack could also indicate a conscientious character. Two visits to small businesses revealed the need for potentially costly alterations to achieve compliance. This could have been prevented but EHPs in both cases understood why people did not approach them earlier, admitted they must get better at communicating environmental health standards and were willing therefore to work with them to achieve compliance.

The history of compliance also provided important clues about the character of the regulated, a poor history resulting in more attention from EHPs until compliance was achieved. People concerned about offending or acting on the advice of EHPs were more likely to elicit the cooperation of EHPs and be given more time for compliance than the 'stubborn people', as some EHPs called them. Before revisiting a fish and chip shop for example Township EHP 2 explained that they were still operating without the correct permits and did not attend the recent (free) hygiene training. On arrival, she expected to serve a spot fine but by the end, having observed good standards and improvements, instead, she wrote a notice listing the outstanding matters and time periods for completion.

The ethnicity, nationality and culture of the regulated

Working with people from 'different' backgrounds to their own, as EHPs phrased it, was something they really enjoyed about their work and they sometimes attributed standards and behaviours to ethnicity and nationality. Whites had a reputation for high standards and expectations and a willingness to complain whereas Black-Africans, particularly those from poorer areas, were less vocal and considered inherently disadvantaged due to apartheid and wider socioeconomic inequalities. Suburban EHP 2 covered areas with large Indian and Coloured populations and whilst the former seemed to complain more, he thought both were trying to resolve their own problems with some success and therefore only contacted him when deadlocked. EHPs also associated ethnicity and nationality with the potential for corruption as covered below.

Inner city EHPs often referred to 'foreigners' or 'immigrants' and associated them with poor environmental health and uncooperative behaviours. Some believed foreigners thought EHPs were targeting them deliberately, even when providing health education, and punishing them only made things worse. EHPs questioned whether they didn't know the law because they had no EHPs in their countries, or they were *organisationally incompetent*, or acting as amoral calculators motivated solely by profits. One manager thought things were more complex:

> What is ok for people from the Congo, is not ok here. What is right for you is not right for me. They don't see why we are making a fuss? It's normal for them, where do we draw the line? Eight people in a room is ok for them, but for us it's overcrowding! We can't be enemies with the people, they are staying here.
>
> (Inner city EH Manager)

EHPs also associated cultural differences with offending, particularly when investigating complaints. For example, Suburban EHP 2 observed that in townships powerful stereos are status symbols but in the suburbs, many people do not like loud music, do not know their neighbours and are more likely to complain. EHPs considered it necessary to investigate all noise complaints but Suburban EHP 2 was sometimes accused by Black-African residents of having a 'bad attitude'. Similarly, in a complaint involving plans for the ritual slaughter of an animal in Urbington Suburban EHP 2 was also divided. On the one hand, such practices were common in rural areas and he had no problems with it, but in Urbington it was an offence unless certain conditions were met and this case is revisited below.

The influence of morality on relations with the regulated

As found in Chapter 5 most environmental health laws incorporate strict liability offences where intent should not be a factor but, as Hutter (1988) also found, questions of moral culpability remained important to EHPs. Cases associated with moral culpability that could result in more punitive approaches included:

- persistent offending
- offences linked to profiteering (e.g. overcrowded homes and crèches, illegal dumping)
- offending outside EHP working hours
- attempts to obstruct and/or deceive EHPs
- knowingly trading without environmental health permits or licenses

As Lipsky (1980) and Maynard-Moody and Musheno (2000, 2003) found, moral overtones frequently shaped the decisions of EHPs, particularly in their views of the regulated. Urbington EHPs were more likely to help those considered worthy, including offenders:

> Let's say you've given a person a statutory notice and he is not complying and you give him maybe another 2 days to rectify the problem, but you have to act, unfortunately. But if you have given a person statutory notices … let's say you've given him 10 requirements and he's done already 7 requirements, give him another chance, he's showing that he's going to complete the statutory notice that you've given him.
>
> (Suburban EHP 2)

Worthy traits included anyone cooperative and trying to improve their environmental health despite lacking knowledge and/or experiencing difficult socioeconomic circumstances and could include the *organisationally incompetent*, *victim* and *fixer* typologies described earlier. Conversely, EHPs were more likely to use punitive approaches towards the less worthy like the uncooperative and those associated with amoral calculator and villain typologies.

EHPs framed their arguments in moral terms for many reasons, not least to persuade people that they were not there to just 'enforce laws for the benefit of the Council' as Township EHP 2 phrased it. She went on to describe how small businesses often accused big businesses of taking their customers, but she asked them to think like a customer:

> Would you rather come to your shop or would you rather go to [the supermarket]? … outside here it's dirty, there's graffiti, you come inside it is dark, not sufficient light, not sufficient ventilation, the walls are black … the ceiling is sagging. The people serving you, you can see that they have not washed. There is no protective clothing. You see the area is a mess. So I'm like those basic things, you are not doing that for the State, you are not doing that for the Environmental Health Practitioner, you are doing that for your business … maybe you don't care about the safety of the food, but just attracting your customers?
>
> (Township EHP 2)

That improving environmental health is good for your health and business was often used by EHPs:

> Here I'm saying: 'put sufficient ventilation'. It means that you must put a window, the reason being that it should provide sufficient ventilation and natural lighting. So the window is going to help you, that during the day you should not switch on the light. At least you will have natural light and it will also allow the cross ventilation of the air from the door to the window, clearing the air and then you will never have a lot of headache …
>
> (Township EHP 1)

> Educate them to make their premises look clean, tell them that by giving people fresh food it's good for their customers as well and it's good for their businesses. Not like … looking for the cockroaches so you can issue them with a fine. We must move away from such things.
>
> (Suburban EHP 2)

Similar moral sentiments also featured across paperwork, including standard paragraphs that offending cost Urbington a lot of money that could be better spent on urgent projects like housing; that all stakeholders, not just the council, were responsible for environmental health; and that by prosecuting offenders, Urbington was protecting peoples' Constitutional rights.

Offenders also used moral arguments to try to persuade EHPs of their worth in the hope of more favourable outcomes. For example, Township EHP 1 measured the floors of all crèches to one decimal place following an inspection where a crèche owner argued, successfully, that once rounded up the total could constitute one more child. One crèche owner used different moral arguments to try to persuade Township EHP 1 to accommodate 10 extra children in her overcrowded crèche. These included accusing Urbington of a 1-year delay with her crèche permit, asking EHP 1 whether she would tell the 10 children and their parents herself that they could no longer attend and accusing EHP 1 and Urbington of targeting her unfairly when there are so many worse crèches nearby. Township EHP 1 acknowledged the delays and the high standards of this crèche but was not prepared to use her discretion to bend the rules.

Returning to the amoral calculators and villains typologies, EHPs sometimes blamed individuals and businesses who should know better:

> If there is a bin in front of you and you prefer to throw on the floor, it's bad behaviour because you know what is the bin for … peoples tend to suffer each other in those wrong doings, we've got shops, they know very well that they have to have a proper waste removal, they are shops. But you find shops dumping at the corner of a street because people are dumping there … they are dumping meat and perishable goods … it's a nuisance …
>
> (Inner city EHP 3)

Accounts of *villains* and *victim*s were also framed in moral terms. Inner city EHP 3 had recently inspected a church sheltering more than 100 migrants including children barely two weeks old. Though conditions in the church were very bad she also questioned how people could have children and then raise them in such difficult circumstances. Certain tasks also created moral tensions, particularly housing cases when the monitoring work of EHPs could result in evictions of the urban poor that they were often expected to attend:

> [Urbington] have the powers to take action, but they want us EHPs to be there on the day of the eviction, of which it is not our role to be evicting the people, or to make sure that the people are eventually evicted.
>
> (Inner City EHP 1)

Indeed, newspaper articles covering these evictions, particularly the plight of the evicted, were often pinned to the walls of the regional offices and it was clear that EHPs felt guilty about their involvement. Blitz work targeting known environmental health hotspots was potentially more confrontational but also justified on moral grounds, including arguments that EHPs were not there to harass people but to secure compliance with the law.

EHPs were also willing to bend the law where there were tensions with the moral aspects of a case. Returning to the complaint about plans for a cow to be slaughtered in Urbington, this had strong moral overtones for Suburban EHP 2. In his rural home, such practices were commonplace but he acknowledged that in the city it was an offence. He was also a qualified meat inspector, arranged to supervise the slaughter and visited the complainants to reassure them of this. This did not resolve their concerns but, much to his relief, the cow was slaughtered elsewhere because, he suspects, of pressure from the complainants and their threats to report the case to his managers and the Society for the Prevention of Cruelty to Animals. The willingness of EHPs to bend the law was further explored using two vignettes based on cases observed:

> *A crèche applying for a permit in your area has been given provisional approval by the Emergency Services, building and licensing inspectors. The person providing the service is friendly and co-operative and is providing facilities and care of a high standard, however your calculations of indoor care area space find that there is only space for 30 children, but the crèche has 35 each day.*

When asked how they would proceed all EHPs began by explaining the law and exploring ways to accommodate the extra children like removing or re-arranging furniture and there was general agreement that five more children were possible, but not many more. They would then issue a written notice of their decision and all agreed that if extra space could not be found, 30 children was the maximum permissible on the permit and appropriate time would be given for compliance. EHPs were unwilling to bend the rules because they sign the permits and wanted to be fair to other crèches. By giving time they also hoped that parents affected would have time to find another crèche. EHPs were also asked if they would act differently if the service provider was aggressive and uncooperative. Most admitted they would act differently in this situation, typically with less flexibility and more punitive approaches, including possible grounds for closure, provided other departments agreed. However, two EHPs claimed they would not act differently because cases should be driven by the facts, responding in kind with aggression could make matters worse.

> *Imagine that you find two identical food premises in your area that have both been trading for 1 month without a Certificate of Acceptability permit. One is owned by a national company, the other is owned by a young family that have taken out a huge loan to set up their new business.*

Most EHPs admitted they would treat both cases the same because the law was the law, it applied to all businesses and trading without a permit was an offence. However, the next step for all was to inspect the premises and start the permit application process. If the premises complies a permit would be issued, if not EHPs would issue a written notice with appropriate timescales for compliance. All agreed this would be easier for the national company but for both premises continued non-compliance would result in more punitive approaches. Township EHPs 1 and 2 added that they were more likely to be lenient towards family shops, partly because of their own failures to educate. Therefore, despite the law attempting to remove morality from offending, it remained integral to the everyday decision-making of EHPs.

Regulating the informal sector

All EHPs considered their work more straightforward in 'developed' areas, as they called them, for reasons including fixed locations and more resources for compliance, but 'developing' areas were a constant challenge:

> I want to do what you say, but I don't have money.
>
> (Inner city EHP 1 imitating an informal business owner)

> Like you have maybe a spaza shop [informal convenience store, often run from home], your tuck shops or maybe your informal crèches, they will be in a shack. So before you even come with the

> By-laws, the By-laws will tell you that's not wanted, that maybe the premises are supposed to be of a solid structure and there are supposed to be hand wash basins, and then double bowl sink, the floor to be of a smooth surface, light coloured paint. You can't enforce that in a shack.
>
> (Township EHP 2)

> But why prosecute people [in informal settlements] if they can't comply with By-Laws? If you're gonna look for something you're gonna find it. It's ridiculous! You can't prosecute them for no water if they have no water! Give them the requirements, but work around them.
>
> (Inner city EHP 4)

Discretion was critical but before taking any action EHPs were careful to establish whether the informal business or settlement was considered 'legal' by Urbington via checks on land ownership and use in liaison with others, particularly Urbington's Departments of Planning and Housing. An illegal decision was referred immediately to the appropriate department and the involvement of EHPs ended there. They knew what could happen next and why informality persisted but EHPs also knew that to provide services before these checks were made risked legitimising what Urbington might consider an 'illegal' occupation. This also helped distance EHPs from what happened next, including forced evictions, though this was not guaranteed and EHPs felt uncomfortable working on these cases.

Further action depended on the case. For example, EHPs monitored conditions in informal settlements and sent reports to the relevant Urbington department in the hope that further action would be taken. Sometimes they were helped by information and complaints from residents but EHPs also expected little feedback for reasons including a general distrust of officials and fears of eviction. EHPs also expected a hostile reception until they got to know people better.

Sometimes EHPs attempted to formalise the informal, like the case where Township EHP 1 eventually persuaded an informal manufacturer of window frames to move to the local industrial zone where his business began to flourish. But EHPs mostly used their discretion to work with informal businesses to achieve minimum safety standards that then enabled these businesses to comply with the law and apply for the appropriate permit. Indeed, Urbington's own compliance data suggested that EHPs found no greater health risks overall for informal food businesses when compared to formal ones, though some EHPs remained uncomfortable about issuing permits to informal food traders:

> So am I going to overlook the fact that the walls will not be of a smooth surface and a light coloured paint? There will obviously be no ceiling, no double wash hand basin, just the hand wash basin … ok can we ignore that and issue a certificate knowing that you've trained this person especially on personal hygiene and how to handle food? … although she's cooking in a shack everything is in shelves, they've tried to put shelves there, they've cleaned this and you feel they are not compromising the consumers, then you issue a certificate.
>
> (Township EHP 2)

Though EHPs were adapting the law to accommodate informal businesses other challenges included illiteracy, with EHPs spending a long time providing verbal advice or physically demonstrating what they wanted to see. EHPs were sometimes also reluctant to attempt more punitive approaches because some people, notably street hawkers, carried no identity documents [an offence in itself in South Africa, but not enforceable by EHPs who needed IDs to complete their own paperwork] and they suspected this was a deliberate tactic. EHPs were also conscious that sanctioning informal businesses could damage local relationships and achieve little in the long term:

> … we just move the problem, not solve it, whilst … everybody flocks to [Urbington]!
>
> (Inner city EH Manager)

Urbington's Street Trading By-laws created further tensions between EHPs and street hawkers. After completing a food hygiene course street hawkers were asked to pay R130 (~£20 in 2007) for a business license and once awarded their (free) food hygiene permit could apply for a trading stand. Having invested time and money the hawkers were often disappointed by the shortage of stands in prime locations and then returned to their old 'illegal' stands. EHPs were very keen to distance themselves from this problem by stressing that they only regulated hygiene, not location (a Police responsibility), but they were often on the receiving end of complaints from local hawkers.

Another problem was that minimum standards for other (non-food) informal premises were still under development. In contrast to the food premises data above, Urbington's compliance data suggested that non-food-related informal businesses identified far more health risks than formal businesses. EHPs also thought these risks would continue with population growth far exceeding the supply of housing and other infrastructure and services but they kept receiving permit applications:

> ... do we actually issue that [informal crèche] with a permit? Maybe yes you could say, because she has to comply with the area, to say how many kids can she put there ... she must provide separate basins for washing ... your basic things. So to be honest I don't know if they're actually going to give them permits ... they will never be fully compliant because of the area, the structure that they find themselves in ...
>
> (Township EHP 2)

The potential for corruption

Frameworks to prevent and deter corruption existed across South African law and the policies of Urbington and the HPCSA but EHPs often commented that corruption was everywhere and cited the 'custom' of handing the Police R20 when they stop you driving in the hope they might waive a higher fine. No EHP considered themselves corrupt and offers of soft drinks or coffee/tea were commonplace, particularly during post-inspection discussions. Most EHPs had a 'can of Coke' or 'R10 rule' to establish clear limits and nearly all offered to pay for any drinks offered immediately. Further, to maintain their distance EHPs typically refused to eat in the food premises they inspected but people were willing to offer all three types of bribe as categorised by Reisman (1979).

For example variance bribes, to stop EHPs from enforcing the law, were the most common and included a wide range of gifts ranging from breakfast cereals to wide-screen TVs if an EHP agreed to overlook offences. One possible transaction bribe, intended to speed up decision-making, was even observed during a visit to a contract caterer. The manager promised to thank the EHP involved 'in the usual way' for fast-tracking his permits by organising a barbecue for his friends and family, an offer this EHP accepted without hesitation.

Attempts at outright purchase bribes, to capture EHPs completely, were recalled by Suburban EHP 2 in the form of covert A4-sized brown envelopes forced upon him by one business owner that really confused him for many reasons, not least because the business was compliant and there seemed so little to gain. EHP 2 refused and explained that there was no fee for the permit but the owner persisted and suggested he buy presents for his girlfriend. EHP 2 went to his car but the owner threw the envelope inside, only for EHP 2 to hand it back stating that his premises comply. The owner tried again and claimed he was not bribing him but merely wanted to say thank you for his quiet and speedy work. The owner also suggested that on future visits EHP 2 should feel free to tell him that the envelope is 'small'. When EHP 2 again refused, the owner instead offered to organise an event for him anytime. As EHP 2 drove away the owner attempted to throw the envelope through his car window and EHP 2 feared someone could be taking pictures to set him up. After the incident, the owner called Suburban EHP 2 to explain how upset and let down he felt. The EHP reported everything to his manager immediately but remained very concerned about it.

The factors EHPs most associated with bribery were ethnicity and nationality. EHPs singled out some Black-African business owners, particularly Nigerians and Congolese, who constantly wanted to, in their words, 'do deals with fellow Africans'. EHPs also singled out Indian-owned businesses and cautioned that they weren't being racist but instead of simply fixing things there seemed to be a culture of negotiation that included 'offers' that EHPs considered bribes.

Lastly, following his experiences in New York, Leff cautioned that EHPs like him were more susceptible to bribery because of poor salaries and warned that:

> ... a city that underpays its municipal employees is likely to be a corrupt city.
>
> (Leff 1988:45)

Observations coincided with reports of an inner city EHP demanding an R2000 variance bribe from the owner of a bad building but he was later caught by colleagues in a sting operation. However, news of bogus EHPs across South Africa was also becoming more common. They typically approached businesses, identified themselves as EHPs using fake IDs, conducted inspections and served spot fines ranging from R1000 to R20000 (~$140–2800 in 2007), with threats of heavier fines if unpaid. One bogus EHP was caught targeting food businesses, drafting notices requiring permits, providing fake application forms and then demanding payment alongside a promise to speed up the process. In response, Urbington were urging people to check the ID cards of EHPs and stressed that many permits were free but payments for others must be made at designated pay points, not to EHPs themselves!

References

Ayres, I. and J. Braithwaite (1992) *Responsive Regulation.* Oxford University Press, Oxford, UK.

Beall, J., O. Crankshaw, and S. Parnell (2000) Victims, villains and fixers: the urban environment and Johannesburg's poor. *Journal of Southern African Studies*, Vol. 26 (4) 833–855.

Hutter, B. (1988) *The Reasonable Arm of the Law? The Law Enforcement Procedures of Environmental Health Officers.* Clarendon Press, Oxford, UK.

Kagan, R. and J. Scholz (1984) The criminology of the corporation and regulatory enforcement strategies, in K. Hawkins and J. Thomas (Eds.) *Enforcing Regulation.* Kluwer-Nijhoff, Boston, USA.

Leff, S. (1988) Corruption in the kitchen. *New York Magazine*, Vol. 21 (41) 38–45.

Lipsky, M. (1980) *Street-level Bureaucracy: Dilemmas of the Individual in Public Services* Russell Sage Foundation, New York, USA.

Maynard-Moody, S. and M. Musheno (2000) State agent or citizen agent: two narratives of discretion. *Journal of Public Administration Research and Theory*, Vol. 10 (2) 329–358.

Maynard-Moody, S. and M. Musheno (2003) *Cops, Teachers, Counsellors: Stories from the Front Lines of Public Services.* University of Michigan Press, USA.

Reisman, M. (1979) *Folded Lies: Bribery, Crusades and Reforms.* The Free Press, London, UK.

8 Conclusions and recommendations

In this final chapter, I bring together my findings and make recommendations to help strengthen and promote the work of local government EHPs in South Africa and beyond.

The local government EHP as a responsive regulator?

At face value my findings provide many examples of EHPs as street-level bureaucrats, intervening in the activities of businesses, the public and others and issuing constant commands to discipline and control the people of Urbington. However, this ignores the complexity of regulation, undersells the considerable potential of EHPs and overlooks where they sometimes go wrong. Instead, the model of governance in Figure 3.2 provides fresh insights into how environmental health inequalities are tackled, or not, by exploring the continuous circulation of power across the streets, workplaces and homes of Urbington. This power was distributed unequally and sometimes not with EHPs themselves, but spaces for challenging it were plentiful and rarely captured or closed down.

Urbington EHPs continued to rely largely on law enforcement to try to tackle environmental health inequalities but the model of governance explains why cases that reach the regulatory pyramid are generally tackled using persuasive approaches. The very existence of more punitive approaches is significant, not least to try to deter environmental health offending, but as Hutter (1988) found in the UK the many uncertainties EHPs have to navigate each day very much framed their responses. I start this section by summarising the main factors influencing the decision-making of EHPs in Table 8.1 before exploring the limits of responsive regulation.

To begin at the base of the regulatory pyramid, Urbington's regulatory context was characterised by huge and persistent inequalities, a

DOI: 10.1201/9781003009931-8

Table 8.1 To punish or persuade: the most influential factors on EHPs

To punish:	
Organisational	• Strategic support – e.g. zero tolerance, targeting offending hotspots
	• Performance targets for spot fines
Individual	• More experienced EHPs
Relations with regulated	• Failure of more persuasive approaches
	• Offenders are considered criminal, *amoral calculators*
	• Serious offences and those affecting the public
	• Uncooperative behaviour of EH offenders
To persuade:	
Regulatory context	• EH inequalities and legacy of apartheid
	• Public distrust and hostility
	• Media criticism of punitive approaches
Legal	• Emphasis on persuasive activities as 'developmental'
	• Uncertain criminality of EH offences
	• 'Low' maximum sanctions risked undermining EHP authority
	• Uncertain legal mandate of EHPs
Organisational	• Expectations of colleagues and managers
	• Performance targets
	• Lack of guidance on punitive approaches
Individual	• EHPs see themselves as educators, not law enforcers
	• Personal safety risks of more punitive approaches
	• Lack of criminal justice system support for prosecuting EH offenders
Relations with regulated	• Most EH offenders are *organisationally incompetent*
	• Informal sector businesses
	• EHPs rely on the cooperation of the regulated

public suspicious of being regulated and a media watching local government EHPs closely and willing to publicise perceived injustices. The legal mandate of EHPs was broad, complex and characterised by uncertainties that gave little encouragement towards the use of more punitive approaches. Urbington's own controls on its EHPs generally favoured persuasion but the views of EHPs themselves were amongst the most influential factors, particularly their desire to be seen as educators and advisors and the very real dangers that punishment could backfire. EHPs were also well aware that they could not rely on the criminal justice system for support in prosecuting offenders. The association of most environmental health offending with *organisational incompetence*, not criminality, was also critical to their preference for persuasion, particularly towards the informal sector. EHPs were also largely reliant on the cooperation of the regulated for reasons including the frequently long-term nature of compliance and their desire to build good working relationships, though the potential for corruption was ever-present.

Towards the top of the regulatory pyramid, the absence of major influences towards more punitive approaches in the regulatory context and the law was no surprise given that the opposite was usually the case. Evidence of Urbington's organisational support for punishment was significant but usually directed towards offending hotspots in the hope of sending a wider message of deterrence to potential offenders. This was increasingly reflected in national policy too, alongside a recognition that punishment was necessary in certain cases and no longer considered non-developmental or linked to apartheid (NDoH 2013; Ngqulunga, 2013). A performance target for spot fines was also significant and whilst EHPs were not unduly concerned about meeting it they remained worried about the signal it sent and its potential to undermine their authority. My results also suggest more experienced EHPs were more willing to use punitive approaches but this was also influenced by a lack of training and guidance. Relations with the regulated were also important, particularly moral culpability and the alignment of a minority of offenders with more traditional criminal behaviours. In Chapter 7 a failure to secure compliance by persuasive approaches and the behaviours of the regulated themselves had considerable influence on EHPs, with offending associated with more traditional criminal behaviours again making punishment more likely.

The responsiveness of Urbington's EHPs was also limited at different stages. Before even reaching the regulatory pyramid their monitoring work for others had mixed success, putting considerable additional pressure on workloads and sometimes undermining other work and local relationships. That EHPs were exploring legal avenues to force other Urbington departments to take responsibility for their environmental health omissions didn't reflect well on the municipality itself, though it illustrates how local democracy can work in unexpected ways.

Limits to responsive regulation were also identified across the governance model. Towards the top of the pyramid, as Crook and Ayee (2006) and Hutter (1988) found, EHPs were sometimes reluctant to use punitive approaches for fear of damaging their relations with the regulated, particularly smaller and informal businesses, and a concern that some sanctions provided a little deterrent for both the poor and the well off. Further, their use could backfire and undermine EHPs' authority, particularly with a criminal justice system under great pressure and unconvinced that environmental health cases were a priority. Towards the base of the pyramid, persuasive approaches were also limited by resources, particularly the limited time EHPs had to provide education, information and advice to businesses and communities. EHPs knew this was a problem and were trying to work around it (e.g. scheduling visits geographically, attending schools, and community meetings) but the scale of challenges made this a constant struggle.

The brief section on the project and promotional pathway in Chapter 3 reflects the limited activity of EHPs here and its secondary importance to the law enforcement pathway. Though there were some fascinating examples of EHPs working with local people to improve environmental health this potential was not yet recognised and recommendations to strengthen this pathway are made below.

Are local government EHPs tackling environmental health inequalities?

The question mark in the title of this book must remain. Though I have tried to capture what EHPs do and identified many factors influencing their effectiveness, measuring success in tackling inequalities remains difficult and continues to be part of the environmental health profession's invisibility problem, both to the wider public health system and to policymakers, politicians etc. (Couch et al. 2016; Dhesi 2019; Hutter, 1988). It would therefore also be incorrect to see the governance model in Figure 3.2 as a prescription for how local government EHPs should work, but the following recommendations are made to try to strengthen their work.

The first is that Urbington reviews its law enforcement pathway, particularly the sanctions at the top of the regulatory pyramid. The continued operation of the spot fine needs review, including whether its scale can be adjusted in a practical and fair way for all and whether it should continue as a performance target. Consideration of the greater use of alternative sanctions like the withdrawal of permits/licenses and the use of prohibitions is also a priority, my findings suggesting they could provide a far more workable deterrent than the spot fine. The role of the criminal justice system should also be included, particularly the judiciary and public prosecutors. It's understandable why environmental health offending remains a low priority for South African courts but following similar experiences in the UK, the training and guidance materials for magistrates and others in environmental health law developed there (e.g. Magistrates Association 2009) could easily be adapted for South Africa.

A second recommendation centres on strengthening the project and promotion pathway. In such a challenging context the potential for activities that mobilise and build the capacity of communities, businesses and wider civil society to address their own environmental health problems cannot be underestimated. Very crudely, how could Urbington and its EHPs better work with those *fixers* already improving their environmental health? Fortunately, these activities were recognised in Urbington's 5-year development plan and EHPs wanted to continue and build on this work, but it remained secondary to law enforcement and some managers were not yet convinced. An investment in more training for EHPs in community mobilization and project management skills would assist, but learning from established organisations with considerable experience and operating locally like the South African Cities Network (SACN 2016) or Slum Dwellers International (SA-SDI 2016) could really help here. Another starting point would be a more systematic review of how EHPs are already supporting civil society and whether there is scope for investing more resources here.

A third recommendation is to review the performance targets of EHPs to better capture the outcomes of their work. A few requirements for data on premises compliance already exist (e.g. numbers of permits issued to formal/informal food premises), but these could be extended across all environmental health services, as EHPs suggested, alongside more comprehensive data on offending trends (e.g. number of permits withdrawn). These activities also need to be better linked to wider developmental outcomes like those identified in Figure 8.1. Such data could also provide more powerful arguments in Urbington and beyond for investing in environmental health services.

A fourth recommendation is to address the gaps in pay and allowances between Urbington and other municipalities and meet the safety concerns and training requests of EHPs. Progress was being made in 2007 but until these fundamentals are addressed it seems likely that high levels of staff turnover will continue. Training budgets were very constrained but the appetite amongst EHPs for further training in both the law enforcement and project & promotion pathways was considerable. The timely emergence of new professional CPD requirements for EHPs to maintain their HPCSA registration to practice also complements this recommendation.

A fifth recommendation is to review what more the HPCSA and SAIEH could be doing to better support, organise and promote their EHP members. In 2007 there was little evidence that the HPCSA or the SAIEH was doing much for EHPs beyond registering them to practice, though more senior EHPs benefited from the SAIEH's bi-annual conference. More recent developments were encouraging and included the publication of the National Environmental Health Policy as a result of the SAIEH's work (Chaka 2013) and the promotion of the SAIEH's former secretary to National Director of Environmental Health. The development of the SAIEH's regional governance and online presence in recent years is also great to see, visit: https://saieh.co.za/. However, Wright et al. warned that the National Director's work remains 'woefully under-capacitated' (2014:20) and unless more is done by all levels to organise and promote EHPs I fear their national voice will remain unheard.

Balancing environmental health and economic interests?

Determining how much EH should be compromised in favour of business and other interests presented a constant challenge. As covered in Chapter 4, the law constantly asks EHPs to consider Constitutional matters (e.g. equality, reasonableness, historic inequalities) in their work but provided little further guidance on how to achieve this and what the right balance might look like. Instead, EHPs used their discretion to determine this and, as Hutter (1988) found, took their cues from many factors including the seriousness of the problem, its history and public impacts, the regulated themselves and the wider context. The governance model in Fig. 3.2, particularly the responsive regulation pyramid, then provided EHPs with the means to act reasonably by accommodating and changing their approaches accordingly.

Amidst all this uncertainty the regulated and wider public had influence over the decision-making of EHPs. The 'balance' of interests could be skewed by bias or corruption but there was no evidence corruption was widespread, nor that EHPs were biased more generally in favour of big business despite some neoliberal influences on their work. Indeed, EHPs spent considerable time supporting small and informal businesses, though at times they could also be accused of criminalizing them. In some cases, EHPs justified this on the grounds that serious offences and/or those affecting the public were being committed. But for others, particularly those making judgements on buildings or settlements that could lead to evictions, EHPs knew they could be making things worse and were keen to distance themselves from the outcomes for some of the city's most disadvantaged residents.

Political influences on regulation

This study also raises important questions about the relationships between EHPs and politicians. On the one hand, I've shown why environmental health regulation remains ideally situated in local government, given its focus on the local and the duties of its officials and elected politicians to be responsive and accountable to the public they serve. However, the potential for local political interference in regulation was ever present and EHPs continued to exercise a great caution here, therefore what alternatives could be considered?

Chapter 2 explained how Urbington's Environmental Health Department already functions as a contractor to its Department of Health client, therefore one alternative is the transfer of Environmental Health into another Urbington-owned utility/agency or outsourcing its work to a company employing EHPs and others (e.g. licensing officers, pest controllers). In both cases, the legal responsibility for delivering environmental health would remain with Urbington and private sector providers might argue that its EHPs would be less burdened by bureaucracy. Concerns about the loss of accountability could be countered by arguments that EHPs are employed as educated and accountable professionals, though two environmental health-related studies from Johannesburg (Samson 2008 and Bond and Dugard 2008) caution that similar arrangements have already enabled two utilities to put profits before peoples' rights. In the UK, similar changes led Tombs (2016) to raise concerns about the loss of local democratic accountability and potential conflicts of interest, alongside the lack of evidence that private providers are actually cheaper or more effective.

Another option could be to move local Environmental Health Services into provincial or national government. This would require changes to the 'municipal health services' clauses across the law but existing provincial or national arrangements for environmental health regulation could be expanded. Local representation could be maintained via the daily work of EHPs in the field, including opportunities for local politicians to contact

EHPs, whilst existing provincial/national government offices could be enlarged to give support and expertise. This option could also enhance the role of the National Director of Environmental Health, which in turn might increase their influence on the national government in this fragmented policy space. But given that local government for all South Africans is less than 30 years old and its environmental health services are younger still and only just becoming established, it is arguably way too early for potentially costly and unproven alternatives like these.

Towards greater visibility and investment: EHPs as doughnut makers?

My findings provide further evidence of Dhesi's (2019) double invisibility problem, where despite environmental health problems being very visible across much of Urbington and South Africa itself the work of its EHPs in response remains invisible within their own organisations and to public health policymakers. The governance model provides new insights into how local government EHPs are trying to tackle upstream environmental health inequalities and the challenges they face but I'm not convinced this is enough to improve their visibility and encourage greater investment. Therefore I propose exploring three new lenses that could help here. The first revisits Raworth's (2017) doughnut economics from Figure 2.1 and how EHPs could be seen as 'doughnut makers' as summarised in Figure 8.1.

Starting with doughnut's inner ring, through both law enforcement and project/promotional activities EHPs are helping to protect and build a stronger social foundation by working across many of the UN's Sustainable Development Goals. They frequently combined their role as Cornell's (1996) 'general practitioners of public health' with specialist knowledge in areas like housing, the informal sector or project management or, if lacking this specialist knowledge, they knew where to find it via colleagues, managers and others like the regulated themselves. As well as monitoring and trying to prevent many shortfalls, my findings also illustrate how EHPs are responding to the voices of businesses, politicians and the wider public, building local relationships and networks to maintain and improve environmental health and being held to account for their actions by many actors. The actions of EHPs in safeguarding the doughnut's outer ring were less extensive but their long-standing work in areas including air and chemical pollution and their emerging work on climate change with local schools and others suggest they are well positioned to strengthen their role in safeguarding the ecological ceiling in the future.

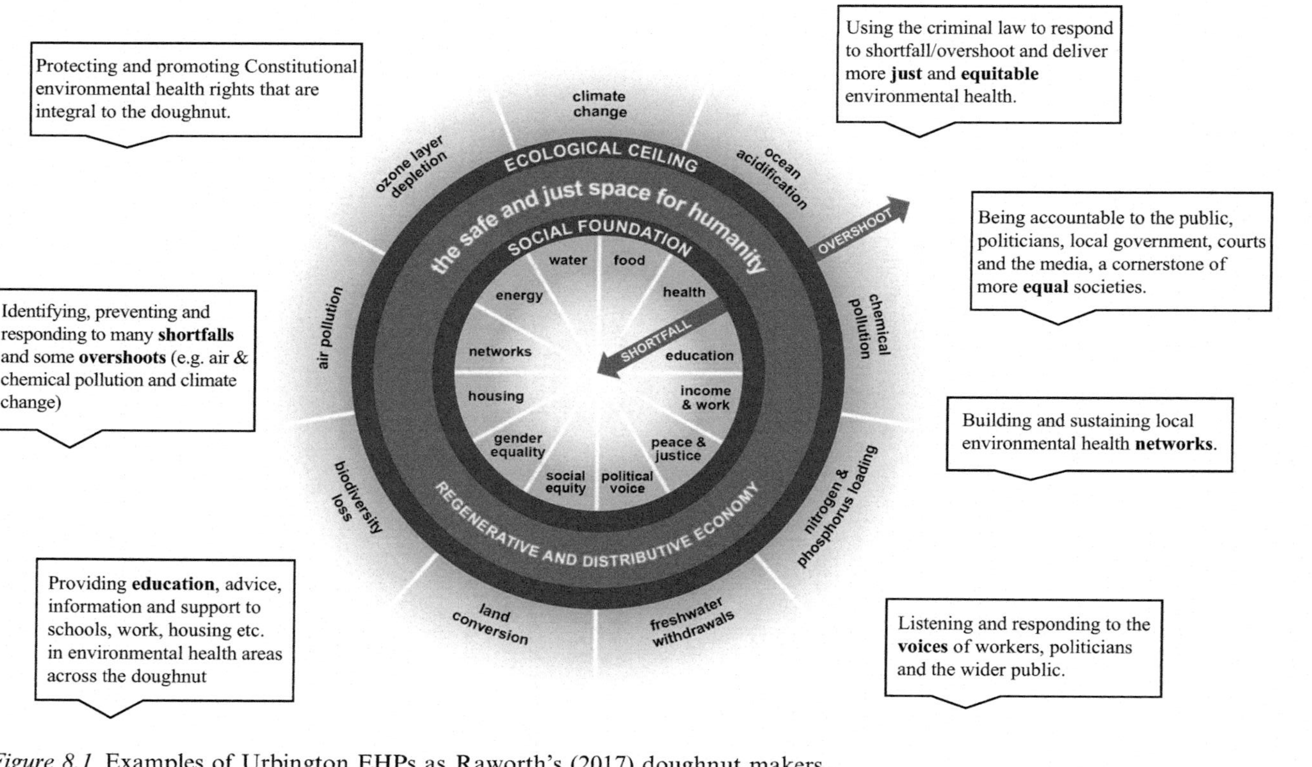

Figure 8.1 Examples of Urbington EHPs as Raworth's (2017) doughnut makers.

Viewing local government EHPs as doughnut makers, therefore, provides a new focused for the relevance of their work in tackling the environmental health inequalities of the 21st century. This argument that EHPs have many of the right ingredients to help create more inclusive, sustainable and better-governed cities echoes Chadwick's (1842/1965) arguments from nearly 200 years ago that improving environmental health was integral to creating better societies. The challenge comes in persuading EHPs to see their work in this way, my findings instead reflecting Hutter and Manning's (1990) comment on regulation as the product of the tensions between senior managers focused on overall performance, middle managers focused on controlling their inspectors and inspectors themselves pre-occupied with their areas. That EHPs at all levels acknowledged their role in delivering Constitutional rights to environmental health provides one example of this vision but those at the street level remained focused on fighting local fires, operations managers were busy supporting and monitoring them and senior managers, crudely, delivering on their key performance indicators. Perhaps doughnut economics could help engage a wider audience and help argue why, as doughnut makers, local government EHPs are worthy of more investment.

Towards greater visibility and investment: EHPs as 21st-century public servants

A second lens looks instead at how EHPs navigate the governance model in Figure 3.2 using many of the skills, attributes and values of Needham and Mangan's (2014) 21st-century public servants. This UK model identifies the characteristics effective public servants will display in the future and how they could be supported to get these skills within a context not dissimilar to South Africa and characterised by cuts to budgets, greater demands for service user voice and increased public expectations. Revisiting my findings using this lens we can see how Urbington's EHPs are:

- *Acting as municipal entrepreneurs* – e.g. developing ways to manage scarce resources, drawing on many techniques (law enforcement, project management) to achieve local improvements and building and navigating systems to improve environmental health across Urbington and other organisations.
- *Acting as co-producers of environmental health* – e.g. creating and sustaining strong and lasting relationships with communities, businesses and others. I would hesitate to describe these as

'partnerships' for many reasons, not least given the boundaries of EHPs as regulators, but the creation and management of these relationships was a form of co-production.

- *Combining generic skills with professional expertise* – e.g. by delivering technical environmental health knowledge using many soft skills including empathy, persuasion (and punishment, where justified), political awareness, the ability to make judgements and to build long-term relationships with a wide range of local communities, businesses and others.
- *Building careers that are fluid across sectors and services* – e.g. by drawing on skills and experience from other councils, provincial and national government, the private sector and others. This breadth of experience helped build relationships and trust within Urbington itself and with the regulated, politicians and others. Most EHPs were also keen to further develop themselves in different ways which included service management experience and further education and training to develop their technical knowledge and soft skills.
- *Combining a public service ethos with commercial skills* – e.g. through motivations to help people, to make a difference and improve local environmental health by drawing on law enforcement and wider skills in project management and promotion and a willingness to work with others including businesses, communities and charities.
- *Surviving with limited resources* – e.g. operating in a context of very limited resources was normal in Urbington and required EHPs to constantly create new ways to work. Discretion was critical, particularly when working to accommodate the informal sector in law enforcement. The development by EHPs themselves of the project and promotion pathway also illustrates their willingness to explore low/no cost, community-based alternative ways to tackle local inequalities.
- *Rooted in place* – e.g. the work of EHPs and their identities were framed by knowledge and pride in their areas and their relations with local businesses, communities and others, including Urbington itself. Despite many tensions and persistent challenges, a sense of serving people and place was strong and shaped the knowledge and daily practice of EHPs.
- *Reflecting on practice and learning from others* – e.g. though EHPs were fighting daily fires their work with colleagues, managers and others provided some time, space and support to reflect upon these skills, attributes and values and to make adjustments when

necessary. The role of supervision was critical here, particularly for less experienced EHPs and when facing complex and/or sensitive cases.

Two of the challenges identified in the 21st-century public servant model were ever present for Urbington's EHPs. The need for fluid and supportive organisations was a constant source of tension. Considerable local discretion and generally supportive managers helped here but to borrow a comment from one of Needham and Mangan's (2014) interviewees, EHPs strived to work as 21st-century public servants from within '19th century organisations' like Urbington that remain hierarchical and controlling. There was also limited evidence of moves towards more collaborative and distributed models of leadership at Urbington. Traditional command and control type relationships remained strong and also influenced relations between EHPs and the regulated, particularly during the law enforcement pathway. At the same time, however, these models were being challenged in many ways including the necessity for EHPs to cross organisational boundaries and work with others as municipal entrepreneurs.

Towards greater visibility and investment: more evidence-based environmental health?

A third and final lens considers my findings as an important contribution towards more evidence-based environmental health. A few years ago, I and some environmental health practitioner-researcher colleagues defined evidence-based environmental health as:

> … environmental health policy and practice supported by the best available evidence, taking into account the preferences of citizens and the wider public and our own professional judgment
>
> (Couch et al. 2016:6)

Implicit in this definition is the ability of EHPs to give a clear and up-to-date rationale for their practice whilst recognising that evidence is often uncertain, changing, vulnerable to politics and can be difficult to access, hence the term 'best available'. But EHPs should have the confidence to embrace these uncertainties and use evidence to improve environmental health. Evidence also works alongside professional judgement because of the limits of the available evidence and the unique and complex nature of cases, whilst judgements should also consider the preferences of all those affected where possible (Couch et al. 2016).

My findings provide important evidence on the work of local government EHPs in tackling environmental health inequalities that could inform policy and practice and be utilized by EHPs and others, particularly those in Universities educating future EHPs. Whilst the evidence-based practice is becoming well established across the wider public health system, for environmental health its influence remains limited (Dhesi 2019). This is also disappointing because EHPs are the epitome of the practitioner-researcher, their daily work involves the collection and interpretation of data, drawing conclusions, making recommendations, communicating these to others and monitoring for improvements. The reasons for the lack of progress here are explored in detail elsewhere (e.g. Couch et al. 2016; Dhesi, 2019) and include poor access to research knowledge, practice-based cultures that do not value research and the politics of public health (Couch et al. 2016). A culture of research and publication amongst students and practising EHPs themselves could help raise the standards and visibility of EHPs within public health, across government and beyond. I hope this rediscovery of the work of local government EHPs and why it matters can contribute towards better environmental health for all in South Africa and beyond.

References

Bond, P. and J. Dugard (2008) The case of Johannesburg water: what really happened at the pre-paid 'Parish pump'. *Law, Democracy and Development*, Vol. 12 (1) 1–28.

Chadwick, E. (1842/1965) *Report on the Sanitary Condition of the Labouring Population of Great Britain 1842 – edited by M W Flinn'*. Edinburgh University Press, Edinburgh, UK.

Chaka, J. (2013) SAIEH President's message. *South African Institute of Environmental Health Website*. Accessed on 1 February 2014 via: https://saieh.co.za

Cornell, S. (1996) Do environmental health officers practice public health? *Public Health*, Vol. 110 73–75.

Couch, R., C. Barratt, S. Dhesi, J. Stewart and A. Page (2016) Chapter 4: Research and evidence based environmental health. *Clay's Handbook of Environmental Health* 21st Edition, Routledge, Abingdon, UK.

Crook, R. and J. Ayee (2006) Urban service partnerships, 'street-level bureaucrats' and environmental sanitation in Kumasi and Accra, Ghana: coping with organisational change in the public bureaucracy. *Development Policy Review*, Vol. 24 (1) 51–73.

Dhesi, S. (2019) *Tacking Health Inequalities: Reinventing the Role of Environmental Health*. Routledge, London, UK.

Hutter, B. (1988) *The Reasonable Arm of the Law? the Law Enforcement Procedures of Environmental Health Officers*. Clarendon Press, Oxford, UK.

Hutter, B. and P. Manning (1990) The contexts of regulation: the impact upon health and safety inspectorates in Britain. *Law & Policy*, Vol. 12 (2) 103–136.

Magistrates Association (2009) *Costing the Earth: Guidance for Sentencers 2009*. Magistrates Association, London, UK.

Needham, C. and C. Mangan (2014) *The 21st Century Public Servant*. University Of Birmingham, UK.

NDoH National Department of Health (2013) *National Environmental Health Policy*. Government Gazette No. 37112 4 December 2013.

Ngqulunga, G. (2013) When all else fails. in *Environmental Health News 2013/14*. Newsletter of the Professional Board for Environmental Health Practitioners of the HPCSA, Pretoria, South Africa.

Raworth, K. (2017) A Doughnut for the Anthropocene: humanity's compass in the 21st century Lancet Planetary Health, Vol. 1 (2).

Samson, M. (2008) Rescaling the state, restructuring social relations – local government transformation in post-apartheid johannesburg and its implications for waste management workers. *International Feminist Journal of Politics*, Vol. 10 (1) 19–39.

SACN South African Cities Network (2016) *State of the Cities Report 2016*. South African Cities Network, Johannesburg, South Africa.

SA-SDI South African Alliance of community organisations and support NGOs affiliated to Shack/Slum Dwellers International (2016) *Brief History of the Alliance* via: https://sasdialliance.org.za/

Tombs, S. (2016) *Social Protection after the Crisis: Regulation without Enforcement*. Policy Press, Bristol, UK.

Wright, C., A. Mathee and M. Oosterhuizen (2014) Challenging times ahead for environmental health in South Africa: the role of the environmental health research network *South African Medical Journal* January 2014, Vol. 104 (1) 20–21.

Index